N. I. Akhiezer

Elements of the Theory of Elliptic Functions

American Mathematical Society
Providence, Rhode Island

Н. И. АХИЕЗЕР

ЭЛЕМЕНТЫ ТЕОРИИ ЭЛЛИПТИЧЕСКИХ ФУНКЦИЙ

«НАУКА», МОСКВА, 1970

Translated from the Russian by H. H. McFaden
Translation edited by Ben Silver

1980 *Mathematics Subject Classification* (1985 *Revision*). Primary 33-01, 33A25; Secondary 14K25, 11F03, 14K20, 30C20, 14K07, 33A30, 32H25, 33A65, 65A05.

ABSTRACT. This book presents a systematic account of the theory of elliptic functions and some of its applications. The main content is intended for engineers who have to work with elliptic functions. Reading this book should not be difficult for people who know the elements of mathematical analysis and the theory of functions in the scope of the first five semesters in the physics-mathematics departments at universities and technical colleges with an advanced program in mathematics.
Bibliography: 29 titles. 24 figures, 27 tables.

Library of Congress Cataloging-in-Publication Data

Akhiezer, N. I. (Naum Il'ich), 1901–
    [Elementy teorii éllipticheskikh funktsii. English]
    Elements of the theory of elliptic functions/N. I. Akhiezer; translated from the Russian by H. H. McFaden; translation edited by Ben Silver.
        p.   cm. — (Translations of mathematical monographs, ISSN 0065-9282; v. 79)
    Translation of: Elementy teorii éllipticheskikh funktsii.
    Includes bibliograpical references.
    ISBN 0-8218-4532-2 (alk. paper)
    1. Functions, Elliptic.   I. Silver, Ben.   II. Title.   III. Series.
QA343.A3813   1990   89-18452
515′.983—dc20   CIP

Information on Copying and Reprinting can be found at the back of this volume.
The paper used in this book is acid-free and falls within the guidelines
established to ensure permanence and durability. ∞
This publication was typeset using $\mathcal{A}_{\mathcal{M}}\mathcal{S}$-TEX,
the American Mathematical Society's TEX macro system.

10 9 8 7 6 5 4 3   02 01 00 99

Table of Contents

Foreword to the Second Russian Edition

The main content of this book, like that of the first edition in 1948, is intended for engineers who have to use elliptic functions.

In preparing the second edition, I improved the original text in several places.

Numerical tables were added.(*) Moreover, a short chapter (the tenth) was added on generalizations of the Tchebycheff polynomials. In it elliptic functions are applied to the solution of some problems in the constructive theory of functions, and hence this chapter is a continuation of the ninth chapter, in which the well-known investigations of P. L. Tchebycheff and E. I. Zolotarev are presented.

Here I want to remember with gratitude my late friend Vsevolod Konstantinovich Baltag, who read through the manuscript of the first edition and made a number of useful remarks.

The author

(*)Borrowed from the Polish translation of [14] (Warsaw, 1963).

CHAPTER 1

General Theorems on Elliptic Functions

§1. On periods of single-valued analytic functions

In all of what follows, in the absence of a statement to the contrary, we understand a function to be a single-valued analytic function whose singularities do not have limit points at a finite distance. If f is such a function and if at each regular point u

$$f(u + \Omega) = f(u),$$

where Ω is a constant, then the number Ω is called a *period* of f. Zero is a trivial period. A function having nontrivial periods is said to be *periodic*.

If $\Omega_1, \ldots, \Omega_n$ are periods of a function f, then for any integers $m_1, \ldots, m_n$ the number $m_1\Omega_1 + \cdots + m_n\Omega_n$ is clearly also a period of it.

If $f(u)$ and $g(u)$ have a period Ω, then the following functions also have the same period:

$$f(u + C), \quad f(u) \pm g(u), \quad f(u)g(u), \quad f(u)/g(u), \quad f'(u).$$

For example, let us prove the last assertion. With this goal we take the function

$$\frac{f(u + h) - f(u)}{h},$$

which in view of the preceding assertions has period Ω. Therefore, at each regular point u

$$\frac{f(u + \Omega + h) - f(u + \Omega)}{h} = \frac{f(u + h) - f(u)}{h}.$$

It now remains to pass to the limit as $h \to 0$.

We prove that for each nonconstant periodic function f there exists a $\mu > 0$ such that every nontrivial period of f satisfies the inequality[1]

$$|\Omega| \geq \mu.$$

---

[1] In this sense it is often said that a nonconstant function cannot have an infinitesimal period.

1

Assuming the contrary, we take nontrivial periods $\Omega_1, \Omega_2, \ldots$ of f such that

$$\lim_{n \to \infty} \Omega_n = 0.$$

Since

$$f(u + \Omega_n) - f(u) = 0$$

for any regular point u of f, and hence

$$\frac{f(u + \Omega_n) - f(u)}{\Omega_n} = 0,$$

it follows that

$$f'(u) = \lim_{n \to \infty} \frac{f(u + \Omega_n) - f(u)}{\Omega_n} = 0$$

at every regular point of f, which implies that f is a constant.

The simplest example of a function with period Ω is $e^{2\pi i u/\Omega}$. Each period of this function has the form $m\Omega$, where m is an integer. Thus, in this case there exists a *primitive* period, namely, Ω. Every other period is an integer multiple of the period Ω. Therefore, this function can be called a *simply periodic* function.

The question arises as to whether there exists a function with $n > 1$ primitive periods. Here n periods are said to be primitive if every period is a linear combination of these periods with integer coefficients and if not every period can be represented as such a combination of fewer fixed periods.

The answer to the question is as follows:

1) *There does not exist a nonconstant function with $n \geq 3$ primitive periods.*

2) *There is a nonconstant function with two given primitive periods if and only if the ratio of these periods is not real.*

The first assertion (with regard to $n \geq 3$ periods) and the negative part of the second assertion (with regard to the case of two periods) form the content of a theorem of Jacobi.

§2. Proof of the Jacobi theorem

The periods of a given function f will be represented as points in the complex plane. Then in any finite part of the plane there are only finitely many of these point-periods, because otherwise they would have a finite limit point, so that there would be a sequence $\{\Omega_k\}_1^\infty$ of periods with a finite limit, and then f would have an infinitesimal period $\Omega_m - \Omega_k$ $(m, k \to \infty)$, which is impossible, since the function is assumed to be nonconstant.

Take a nontrivial period Ω and consider the periods $m\Omega$ ($m = \pm 1$, $\pm 2, \ldots$); they lie on some straight line Z. Two cases are conceivable a priori: 1) all the periods of f lie on Z; 2) not all the periods of f lie on Z.

Let us analyze the first case. Since the segment of Z from $-\Omega$ to $+\Omega$ contains only finitely many point-periods, there is a nontrivial period with smallest modulus, and we can assume without loss of generality that Ω is precisely this period. Since all the periods lie on Z, every period can be represented in the form $t\Omega$, where t is real; furthermore, t satisfies the inequality $|t| \geq 1$, since Ω is a nontrivial period with smallest modulus. We prove that t runs through only integer values. This will imply that Ω is a primitive period, and f is a simply periodic function.

Let $t = m + r$, where m is an integer and $0 \leq r < 1$. Since not only $t\Omega$ but also $m\Omega$ is a period of f, it follows that $r\Omega = t\Omega - m\Omega$ is also a period, and this, as we established, is impossible if $0 < r < 1$. Consequently, $r = 0$, i.e., t is an integer.

We proceed to the second case. Suppose that not all the periods of f lie on Z. Denote by Ω' one of the periods not on Z, and consider the triangle with vertices 0, Ω, and Ω'. By what was proved, only finitely many point-periods can lie inside and on the boundary of this triangle. Taking instead of one of the nonzero vertices of our triangle some point-period lying inside (or on a side), we get an analogous triangle containing fewer point-periods. Continuing this reduction, we arrive at a triangle with no point-periods inside it or on its sides, except for the vertices. Without loss of generality we can assume that this "empty" triangle is the original triangle with vertices 0, Ω, and Ω'. We now construct the parallelogram with vertices 0, Ω, $\Omega+\Omega'$, Ω' (Figure 1). The empty triangle with vertices 0, Ω, and Ω' considered earlier represents the "left" half of this parallelogram. We assert that the "right" half of the parallelogram also is an empty triangle, i.e., does not contain point-periods, neither inside nor on its sides (other than the vertices). Indeed, if the right half contained a point-period $\widetilde{\Omega}_1$, then the left half would contain the point-period $\Omega + \Omega' - \widetilde{\Omega}_1 = \widetilde{\Omega}_2$; but the left half is empty by construction. Accordingly, the parallelogram is empty. We now take some period Ω^* of our function. It has a unique representation in the form $\Omega^* = t\Omega + t'\Omega'$, where t and t' are real numbers. This representation is equivalent to decomposing the vector Ω^* with respect to the vectors Ω and Ω'.

If we prove that t and t' are integers, then it will be proved that in the second case of our alternative the number of primitive periods is equal to

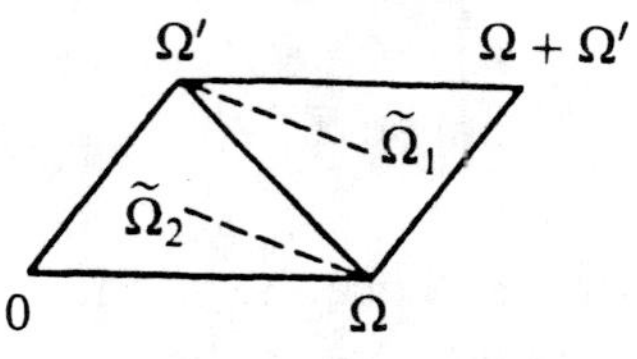

FIGURE 1

two, and their ratio is not real. The proof of Jacobi's theorem will thereby be complete.

Thus, let $t = m + r$ and $t' = m' + r'$, where m and m' are integers and $0 \le r, r' < 1$. We must prove that $r = r' = 0$.

Since $m\Omega$ and $m'\Omega'$ are periods of f,

$$\Omega_1^* = \Omega^* - m\Omega - m'\Omega' = r\Omega + r'\Omega'$$

is also a period. The point-period Ω_1^* lies in the parallelogram with vertices $0, \Omega, \Omega + \Omega', \Omega'$, and hence it must coincide with one of the vertices, since the parallelogram is empty. Thus, each of the numbers r and r' must equal 0 or 1. Since $0 \le r, r' < 1$, it follows that $r = 0$ and $r' = 0$, as was required.

§3. Theta functions

Trigonometric series provide excellent examples of periodic functions. We consider here trigonometric series defining the so-called *theta functions*. As the basic theta functions we take

$$\vartheta_3(v) = \vartheta_3(v|\tau) = \sum_{m=-\infty}^{\infty} e^{(m^2\tau + 2mv)\pi i}. \tag{1}$$

Here v is the argument, and τ is a parameter whose imaginary part is positive: $\Im\tau > 0$. Under this condition the quantity $h = e^{\pi i\tau}$ is less than 1 in modulus, which implies that the series converges absolutely for any finite v.

It is not hard to reduce $\vartheta_3(v)$ to the form

$$\vartheta_3(v) = 1 + 2h\cos 2\pi v + 2h^4 \cos 4\pi v + 2h^9 \cos 6\pi v + \cdots.$$

Thus, $\vartheta_3(v)$ is an entire function of v with period 1.

On the basis of (1),

$$\vartheta_3(v + \tau) = \sum_{m=-\infty}^{\infty} e^{(m^2\tau + 2mv + 2m\tau)\pi i}$$

$$= e^{-\pi i(\tau + 2v)} \sum_{m=-\infty}^{\infty} e^{[(m+1)^2\tau + 2(m+1)v]\pi i}$$

$$= e^{-\pi i(\tau + 2v)} \sum_{n=-\infty}^{\infty} e^{(n^2\tau + 2nv)\pi i}.$$

We see, consequently, that

$$\vartheta_3(v + \tau) = e^{-\pi i(2v + \tau)}\vartheta_3(v). \tag{2}$$

Let us take the logarithms of both sides of (2) and then take the second derivative with respect to v of both sides. We get

$$\frac{d^2}{dv^2}\ln\vartheta_3(v + \tau) = \frac{d^2}{dv^2}\ln\vartheta_3(v).$$

Since, moreover,

$$\frac{d^2}{dv^2}\ln\vartheta_3(v + 1) = \frac{d^2}{dv^2}\ln\vartheta_3(v),$$

it follows that

$$\varphi(v) = \frac{d^2}{dv^2}\ln\vartheta_3(v)$$

is an example of a function with periods 1 and τ, the ratio of which is not real; this function is *doubly periodic*. We remark that $\varphi(v)$ is a meromorphic function, and all its poles have multiplicity two. Indeed, $\vartheta_3(v)$ is an entire function; hence the only singularities of its logarithmic derivative are simple poles, which coincide with the zeros of $\vartheta_3(v)$, and thus the only singularities of $\varphi(v)$ are poles of order 2.

In addition to $\vartheta_3(v)$ we introduce three more theta functions:

$$\vartheta_k(v) = \vartheta_k(v|\tau) \qquad (k = 0, 1, 2).$$

They can be defined by the equalities

$$\vartheta_0(v) = \vartheta_3(v + 1/2),$$
$$\vartheta_1(v) = ie^{-\pi i(v - \tau/4)}\vartheta_3(v + (1 - \tau)/2),$$
$$\vartheta_2(v) = e^{-\pi i(v - \tau/4)}\vartheta_3(v - \tau/2).$$

By using these definitions it is not hard to obtain expansions of all theta functions in Fourier series, as well as to derive theta function reduction formulas that recall well-known formulas in trigonometry. All this is contained in Table VIII (Appendix I).

Concluding this section, we note that the ratios

$$\varphi_1(v) = \frac{\vartheta_1(v)}{\vartheta_0(v)}, \qquad \varphi_2(v) = \frac{\vartheta_2(v)}{\vartheta_0(v)}, \qquad \varphi_3(v) = \frac{\vartheta_3(v)}{\vartheta_0(v)},$$

as follows from the formulas in Table VIII, satisfy the equalities

$$\begin{aligned}
\varphi_1(v+1) &= -\varphi_1(v), & \varphi_1(v+\tau) &= \varphi_1(v), \\
\varphi_2(v+1) &= -\varphi_2(v), & \varphi_2(v+\tau) &= -\varphi_2(v), \\
\varphi_3(v+1) &= \varphi_3(v), & \varphi_3(v+\tau) &= -\varphi_3(v).
\end{aligned}$$

Therefore, the function $\varphi_1(v)$ has periods 2 and τ, $\varphi_2(v)$ has periods 2 and $1 + \tau$, and, finally, $\varphi_3(v)$ has periods 1 and 2τ. Each function $\varphi_k(v)$ is a meromorphic function. Thus, we have a second proof of the existence of doubly periodic meromorphic functions.

Doubly periodic meromorphic functions bear the name *elliptic functions*. The origin of this term will be explained below.

§4. Liouville's theorems

We consider elliptic functions with primitive periods Ω and Ω', and we agree to assume that the ratio $\tau = \Omega'/\Omega$ has positive imaginary part unless otherwise stated. In the complex plane we take a point c and construct the parallelogram with vertices[2] c, $c + \Omega$, $c + \Omega + \Omega'$, $c + \Omega'$. Of the four vertices we include in the parallelogram only the vertex c, and of the four sides we include only the ones meeting at c. The resulting point set is called a *period parallelogram*. We say that two points u' and u'' are *congruent modulo the periods Ω and Ω'*, or *equivalent*, if

$$u'' - u' = m\Omega + m'\Omega',$$

where m and m' are integers, and in this case we write

$$u'' \equiv u' \quad (\mathrm{mod}(\Omega, \Omega')).$$

In view of our definition, a period parallelogram does not contain a pair of equivalent points. On the other hand, for any point u there is a point in a period parallelogram equivalent to it, and, of course, it is unique. Indeed, there are real numbers t and t' such that

$$u - c = t\Omega + t'\Omega'.$$

Setting $t = m + r$ and $t' = m' + r'$, where m and m' are integers and $0 \le r$, $r' < 1$, we find that

$$u - (c + r\Omega + r'\Omega') = m\Omega + m'\Omega'.$$

____

[2]This passage from vertex to vertex corresponds to a circuit of the boundary of the parallelogram in the positive direction, since $\mathfrak{I}(\Omega'/\Omega) > 0$.

Consequently, u is equivalent to the point $c + r\Omega + r'\Omega'$, which belongs to the period parallelogram.

In the study of an elliptic function we can confine ourselves to any period parallelogram.

The initial vertex c of the parallelogram is arbitrary. Due to this arbitrariness we can construct a period parallelogram in such a way that the function does not take some predetermined values on its sides (for example, does not become infinite). Such a choice of the period parallelogram is possible because an elliptic function, like every meromorphic function, takes each of its values only finitely many times in a finite region.

All points mutually congruent modulo the periods form (as is commonly said) a *regular system* or *network* of points on the plane. Corresponding to each such system is a network of parallelograms that fit together to cover the whole plane.

Let $f(u)$ be an elliptic function with primitive periods Ω and Ω', and let the period parallelogram be chosen so that $f(u)$ is regular on its sides.

We integrate $f(u)$ along the contour of the parallelogram. By Cauchy's theorem, the result of the integration is the sum of the residues of $f(u)$ with respect to all poles inside the parallelogram, multiplied by $2\pi i$. On the other hand,

$$\int_{\square} f(u)\, du = \int_{c}^{c+\Omega} f(u)\, du + \int_{c+\Omega}^{c+\Omega+\Omega'} f(u)\, du$$
$$+ \int_{c+\Omega+\Omega'}^{c+\Omega'} f(u)\, du + \int_{c+\Omega'}^{c} f(u)\, du.$$

Making the substitution $u = v + \Omega'$ in the third integral on the right-hand side, we get that

$$\int_{c+\Omega+\Omega'}^{c+\Omega'} f(u)\, du = \int_{c+\Omega}^{c} f(v + \Omega')\, dv = \int_{c+\Omega}^{c} f(v)\, dv,$$

since $f(v + \Omega') = f(v)$.

Consequently, the third integral cancels with the first. Similarly, the second and fourth integrals cancel each other. Accordingly,

$$\int_{\square} f(u)\, du = 0,$$

and hence the sum of the residues of $f(u)$ with respect to all the poles inside this parallelogram is equal to zero. In view of our definition, of two parallel sides only one can belong to the period parallelogram. Therefore, this result is valid also when the function has poles on the boundary of the parallelogram. It is only necessary to take all the poles lying *in* the parallelogram (and not only *inside* it).

Our result has important consequences. Proceeding to them, we define the *a-points* of a function to be the points at which the function takes the value a.

1°. Taking instead of the elliptic function $f(u)$ the elliptic function

$$\varphi(u) = \frac{f'(u)}{f(u) - a},$$

where a is a constant, we find that *the properly counted* (i.e., counted with multiplicity taken into account) *number of poles of a nonconstant elliptic function $f(u)$ in a period parallelogram is equal to the properly counted number of a-points, for an arbitrary a.*

2°. *There does not exist a nonconstant elliptic function that is regular in a period parallelogram.* Indeed, the number of poles of such a function would be equal to zero, and hence so would the number of a-points (by the preceding assertion), for an arbitrary a, which is absurd.[3]

3°. *The number of poles of an elliptic function in a period parallelogram, counting multiplicity, cannot be less than two* (this number is called the *order* of an elliptic function).

Thus, two types of elementary elliptic functions are conceivable a priori: A function of the first type has in a period parallelogram one pole of second order, with residue equal to zero; a function of the second type has two distinct poles of first order with residues differing only in sign. Functions of both types will be constructed below.

All these propositions bear the name of *Liouville's theorems*. Another theorem is due to him. Proceeding to it, we denote by $\alpha_1, \ldots, \alpha_m$ the a-points of $f(u)$ lying in a period parallelogram, with each point written as many times as its multiplicity has units. Further, let $\beta_1, \ldots, \beta_m$ denote the poles of the function, written according to the same principle. Here it is assumed, of course, that $f(u)$ is not a constant.

By Cauchy's theorem,

$$2\pi i \left\{ \sum_{k=1}^{m} \alpha_k - \sum_{k=1}^{m} \beta_k \right\} = \int_{\square} u \frac{f'(u)}{f(u) - a}\, du,$$

where it is assumed that on the contour of the parallelogram under consideration $f(u)$ does not take the value a and does not have poles. Computing

---

[3] Proposition 2° follows also from the fact that a periodic function regular in some parallelogram is also regular and bounded in the whole open plane, by periodicity.

the contour integral as above, we find that

$$\int_{\square} u\frac{f'(u)}{f(u) - a}\,du = \int_{c}^{c+\Omega} u\frac{f'(u)}{f(u) - a}\,du + \int_{c+\Omega}^{c+\Omega+\Omega'} + \int_{c+\Omega+\Omega'}^{c+\Omega'} + \int_{c+\Omega'}^{c}$$
$$= J_1 + J_2 + J_3 + J_4.$$

Making the substitution $u = v + \Omega'$ in J_3, we get that

$$J_3 = \int_{c+\Omega}^{c} (v + \Omega')\frac{f'(v)}{f(v) - a}\,dv.$$

Therefore,

$$J_1 + J_3 = \Omega'\int_{c+\Omega}^{c} \frac{f'(v)}{f(v) - a}\,dv$$
$$= \Omega'\{\ln[f(c) - a] - \ln[f(c + \Omega) - a]\};$$

and since $f(c) - a = f(c + \Omega) - a$, it follows that $\ln[f(c) - a]$ differs from $\ln[f(c + \Omega) - a]$ only by an integer multiple of $2\pi i$; hence,

$$J_1 + J_3 = \Omega' \cdot 2n'\pi i.$$

It can be proved similarly that

$$J_2 + J_4 = \Omega \cdot 2n\pi i.$$

Consequently,

$$\sum_{k=1}^{m} \alpha_k - \sum_{k=1}^{m} \beta_k = n\Omega + n'\Omega',$$

i.e.,

4°. *The sum of the a-points of $f(u)$ for an arbitrary a is congruent modulo the periods to the sum of the poles of the function if all the a-points and poles in a single period parallelogram are being considered.*

§5. The Weierstrass function $\wp(u)$

We consider the series[4]

$$\sideset{}{'}\sum_{m,m'} \frac{1}{|2m\omega + 2m'\omega'|^p}, \tag{1}$$

where the summation is over all integers m and m' except for the pair[5] $m = m' = 0$, and the numbers ω and ω' satisfy the assumption made

---

[4]Beginning with this section we often introduce the factor 2 in the notation for periods (setting $\Omega = 2\omega$ and $\Omega' = 2\omega'$).

[5]This is indicated by the prime after the summation sign.

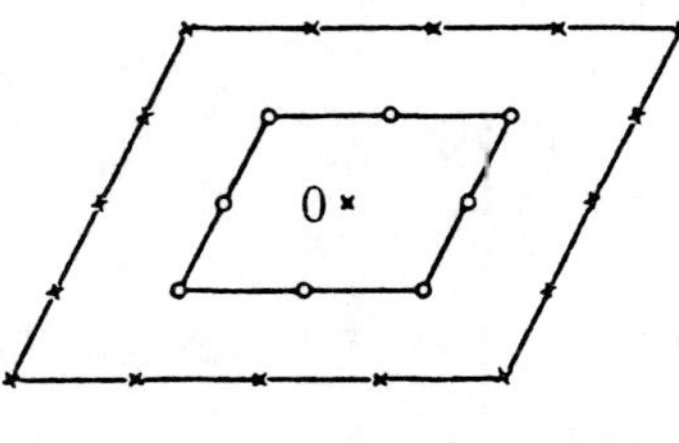

$$\textsc{Figure } 2$$

above. Let us prove that the series converges for $p > 2$ and diverges for $p \leq 2$.

We have here some regular system of points $2m\omega + 2m'\omega'$, from which the point 0 is removed. Let us first of all take the points

$$\pm 2\omega, \quad \pm(2\omega + 2\omega'), \quad \pm 2\omega', \quad \pm(2\omega - 2\omega') \tag{2}$$

of our regular system. These points are the vertices of four parallelograms coming together at the point 0 (Figure 2), and they form the first framing of the point 0. Suppose that the minimal distance from 0 to a vertex of the system (2) is d, and the maximal distance is D. Then the sum of the eight terms of the series (1) corresponding to the vertices (2) satisfies the inequalities

$$8/D^p \leq S_1 \leq 8/d^p.$$

We now take the vertices of our regular system that belong to the second framing of 0. There will be 16 of these vertices, and the minimal and maximal distances from 0 to them will be $2d$ and $2D$, respectively. Therefore, the sum S_2 in the series (1) corresponding to these 16 vertices satisfies the inequalities

$$16/(2D)^p \leq S_2 \leq 16/(2d)^p;$$

the nth framing will consist of $8n$ vertices, and the sum S_n corresponding to it satisfies the inequalities

$$8n/(nD)^p \leq S_n \leq 8n/(nd)^p.$$

Convergence of our series (1) is equivalent to convergence of the series $S_1 + S_2 + \cdots$, and our assertion is an immediate consequence of the fact that

$$S_n \leq \frac{8}{d^p n^{p-1}} \quad \text{and} \quad S_n \geq \frac{8}{D^p n^{p-1}}.$$

In view of what has been proved, the series

$$\sum_{m,m'} \frac{1}{(u - 2m\omega - 2m'\omega')^3} \tag{3}$$

converges absolutely and uniformly in each bounded region of the u-plane if the finite number of terms that become infinite there are removed. Therefore, the sum of the series (3) is a meromorphic function whose only poles (which have order three) are the points $2m\omega + 2m'\omega'$.

Let

$$Q(u) = -2 \sum_{m,m'} \frac{1}{(u - 2m\omega - 2m'\omega')^3}.$$

We show that this function has periods 2ω and $2\omega'$. Indeed,

$$Q(u + 2\omega) = -2 \sum_{m,m'} \frac{1}{(u + 2\omega - 2m\omega - 2m'\omega')^3}.$$

Setting $m - 1 = n$, we rewrite this formula in the form

$$Q(u + 2\omega) = -2 \sum_{n,m'} \frac{1}{(u - 2n\omega - 2m'\omega')^3}.$$

But since the pair (n, m') runs through the same collection as the pair (m, m'), it follows that the right-hand side of the formula is $Q(u)$, and the equality $Q(u + 2\omega) = Q(u)$ is proved. It is proved in exactly the same way that $Q(u + 2\omega') = Q(u)$.

Analogous arguments show that $Q(u)$ is an odd function. Indeed,

$$Q(-u) = 2 \sum_{m,m'} \frac{1}{(u + 2m\omega + 2m'\omega')^3}$$

$$= 2 \sum_{n,n'} \frac{1}{(u - 2n\omega - 2n'\omega')^3} = -Q(u).$$

Here we take into account that the pairs (m, m') and (n, n') with $n = -m$ and $n' = -m'$ run through one and the same collection.

By integration we now introduce the function

$$\wp(u) = \frac{1}{u^2} + \int_0^u \left\{ Q(u) + \frac{2}{u^3} \right\} du.$$

Here it is assumed that the path of integration does not go through vertices in the period network different from the point $u = 0$. Thus,

$$\wp'(u) = Q(u), \tag{4}$$

and, on the other hand, termwise integration give us that

$$\wp(u) = \frac{1}{u^2} + {\sum_{m,m'}}' \left\{ \frac{1}{(u - 2m\omega - 2m'\omega')^2} - \frac{1}{(2m\omega + 2m'\omega')^2} \right\}. \tag{5}$$

Since $Q(u)$ is an odd function, it follows that $\wp(u)$ is an even function. This circumstance can also be obtained easily with the help of the representation (5).

Further, since $Q(u)$ has period 2ω, (4) implies that

$$\wp'(u + 2\omega) = \wp'(u),$$

and hence

$$\wp(u + 2\omega) = \wp(u) + c, \tag{6}$$

where c is a constant.

It follows from the expansion (5) that the only poles of $\wp(u)$ are the points $2m\omega + 2m'\omega'$; therefore, $\wp(u)$ is finite at the points ω and ω'. But since the substitution $u = -\omega$ in (6) gives

$$\wp(\omega) = \wp(-\omega) + c,$$

it follows from the evenness of $\wp(u)$ that c has the value 0, i.e., $\wp(u+2\omega) = \wp(u)$. It can be verified similarly that $\wp(u + 2\omega') = \wp(u)$. We see that $\wp(u)$ is an elliptic function of second order, since it has a total of one pole of order 2 in each period parallelogram. Thus, $\wp(u)$ is one of the elementary elliptic functions in the sense ascribed to this word in §4. It is basic in the theory of Weierstrass.

§6. The differential equation of the function $\wp(u)$

In a neighborhood of the point $u = 0$ the function $\wp(u)$ has the form[6]

$$\wp(u) = \frac{1}{u^2} + 3u^2 {\sum_{m,m'}}' \frac{1}{(2m\omega + 2m'\omega')^4}$$

$$+ 5u^4 {\sum_{m,m'}}' \frac{1}{(2m\omega + 2m'\omega')^6} + \cdots .$$

We adopt the notation

$${\sum_{m,m'}}' \frac{1}{(2m\omega + 2m'\omega')^4} = \frac{g_2}{60}, \qquad {\sum_{m,m'}}' \frac{1}{(2m\omega + 2m'\omega')^6} = \frac{g_3}{140}.$$

In this notation

$$\wp(u) = \frac{1}{u^2} + \frac{g_2}{20}u^2 + \frac{g_3}{28}u^4 + \cdots . \tag{1}$$

---

[6]The rest of the coefficients (to within some numerical factors) are equal to the series

$${\sum_{m,m'}}' \frac{1}{(2m\omega + 2m'\omega')^{2l}} \qquad (l = 4, 5, 6, \dots).$$

We encounter them again in §10.

From this,

$$\wp'(u) = -\frac{2}{u^3} + \frac{g_2}{10}u + \frac{g_3}{7}u^3 + \cdots.$$

Therefore,

$$[\wp'(u)]^2 = \frac{4}{u^6}\left\{1 - \frac{g_2}{10}u^4 - \frac{g_3}{7}u^6 + \cdots\right\},$$

$$[\wp(u)]^3 = \frac{1}{u^6}\left\{1 + \frac{3g_2}{20}u^4 + \frac{3g_3}{28}u^6 + \cdots\right\}.$$

In view of these expansions and (1),

$$[\wp'(u)]^2 - 4[\wp(u)]^3 + g_2\wp(u) = -g_3 + Au^2 + Bu^4 + \cdots.$$

The left-hand side is an elliptic function with periods 2ω and $2\omega'$. Only the points $2m\omega + 2m'\omega'$ can be poles of it. But since this function is regular and equal to $-g_3$ at the point $u = 0$ (as the above formula shows), it is regular in every period parallelogram with $u = 0$ as an interior point, and hence it is a constant by Liouville's theorem. Accordingly, we have obtained the relation

$$[\wp'(u)]^2 = 4[\wp(u)]^3 - g_2\wp(u) - g_3. \tag{2}$$

In other words, $\wp(u)$ satisfies the differential equation

$$z'^2 = 4z^3 - g_2 z - g_3.$$

Equation (2) enables us to express all the derivatives of $\wp(u)$ in terms of $\wp(u)$ and $\wp'(u)$; for example,

$$\wp'' = 6\wp^2 - \tfrac{1}{2}g_2,$$
$$\wp''' = 12\wp\wp',$$
$$\wp^{(\mathrm{IV})} = 120\wp^3 - 18g_2\wp - 12g_3.$$

Let

$$4z^3 - g_2 z - g_3 = 4(z - e_1)(z - e_2)(z - e_3).$$

Then

$$e_1 + e_2 + e_3 = 0, \tag{3}$$
$$e_1 e_2 + e_2 e_3 + e_3 e_1 = -\tfrac{1}{4}g_2, \tag{4}$$
$$e_1 e_2 e_3 = \tfrac{1}{4}g_3. \tag{5}$$

By (3) and (4),

$$e_1^2 + e_2^2 + e_3^2 = \tfrac{1}{2}g_2.$$

Observing that $\wp'(u)$ is an odd function and setting $u = -\omega$ in the equality

$$\wp'(u + 2\omega) = \wp'(u),$$

we find that $\wp'(\omega) = 0$ (it should be kept in mind that $\wp'(\omega)$ is finite). It can be shown similarly that

$$\wp'(\omega') = 0, \qquad \wp'(\omega + \omega') = 0.$$

We see that the points ω, $\omega + \omega'$, and ω' are zeros of $\wp'(u)$, and simple zeros as well, since $\wp'(u)$ is an elliptic function of third order.

We now observe that the quantities $\wp(\omega)$, $\wp(\omega + \omega')$, and $\wp(\omega')$ are all distinct. Indeed, if, for example, $\wp(\omega) = \wp(\omega + \omega')$, then the second-order elliptic function $\wp(u) - \wp(\omega)$ would have the two second-order zeros ω and $\omega + \omega'$, which is impossible.

In view of (2) the quantities $\wp(\omega)$, $\wp(\omega + \omega')$, and $\wp(\omega')$ coincide with the roots of the polynomial $4z^3 - g_2 z - g_3$. Therefore, the numbers e_1, e_2, and e_3 are all distinct.

It is frequently more convenient to use the notation

$$2\omega_1 = 2\omega,$$
$$2\omega_2 = -2\omega - 2\omega' \qquad (\tau = \omega_3/\omega_1, \ \Im\tau > 0),$$
$$2\omega_3 = 2\omega',$$

so that $\omega_1 + \omega_2 + \omega_3 = 0$. Further, one sets

$$\wp(\omega_k) = e_k \qquad (k = 1, 2, 3).$$

It is useful to remark that formulas (3)–(5) lead to the following representation of the *discriminant*:

$$g_2^3 - 27g_3^2 = 16(e_1 - e_2)^2(e_2 - e_3)^2(e_3 - e_1)^2,$$

as well as the equality

$$\tfrac{3}{2}g_2 = (e_1 - e_2)^2 + (e_2 - e_3)^2 + (e_3 - e_1)^2.$$

From this it follows at once that

$$J \equiv \frac{g_2^3}{g_2^3 - 27g_3^2} = \frac{[(e_1 - e_2)^2 + (e_2 - e_3)^2 + (e_3 - e_1)^2]^3}{2 \cdot 27 \cdot (e_1 - e_2)^2(e_2 - e_3)^2(e_3 - e_1)^2}. \tag{6}$$

This formula will be needed in what follows.

CHAPTER 2

Modular Functions

§7. Invariants

In what follows we need to consider rational functions of x and $\sqrt{\varphi(x)}$, where

$$\varphi(x) = a_0 x^4 + 4a_1 x^3 + 6a_2 x^2 + 4a_3 x + a_4 \qquad (1)$$

is an arbitrary fourth-degree polynomial without multiple roots.

The coefficient a_0 can be equal to zero. If this holds, then one of the roots of the polynomial (1) "has withdrawn" to infinity. If $a_0 = 0$, then the coefficient a_1 will be assumed to be nonzero. This need not have been stipulated, since the vanishing of both the coefficients a_0 and a_1 means the existence of a multiple root of (1) at infinity.

If x is subjected to the linear fractional transformation

$$x = \frac{\alpha y + \beta}{\gamma y + \delta} \qquad \left(D = \begin{vmatrix} \alpha & \beta \\ \gamma & \delta \end{vmatrix} \neq 0 \right), \qquad (2)$$

then the rational function becomes a certain rational function of y and of the square root of some new polynomial

$$\psi(y) = b_0 y^4 + 4b_1 y^3 + 6b_2 y^2 + 4b_3 y + b_4.$$

Indeed,

$$R_1(x, \sqrt{\varphi(x)}) = R_1 \left(\frac{\alpha y + \beta}{\gamma y + \delta}, \frac{\sqrt{\psi(y)}}{(\gamma y + \delta)^2} \right) = R_2(y, \sqrt{\psi(y)}).$$

It is natural to deal with the coefficients of the transformation (2) in such a way that the new polynomial has an especially convenient form for the required considerations. One such form is the canonical form

$$\psi(y) = 4y^3 - g_2 y - g_3, \qquad (3)$$

which we encountered in §6 and which was introduced by Weierstrass in the theory of elliptic functions.

15

The form (3) is characterized first of all by the absence of the fourth and second powers of the independent variable. We can see without the detailed computations that there is a transformation (2) that reduces to such a form. Indeed, if we take some root of $\varphi(x)$, say c, and let $x = c + 1/z$, then we get that

$$\varphi(x) = \frac{\varphi'(c)z^3 + \frac{1}{2}\varphi''(c)z^2 + \cdots}{z^4},$$

where $\varphi'(c) \neq 0$, since $\varphi(x)$ does not have multiple roots. If we further let $z = Ay + B$, then for a suitable choice of B the coefficient of y^2 will be zero. Hence, by suitably choosing A, we can get the coefficient of y^3 to be 4, i.e., the polynomial $\psi(y)$ will have the form (3).

The so-called *invariants* are useful for obtaining the final expressions for the new coefficients in terms of the old ones. We give the definition of invariants a little later, after the present considerations.

For convenience we proceed from polynomials of a single variable to homogeneous functions, i.e., forms, in two variables:

$$\varphi(x) = \varphi\left(\frac{x_1}{x_2}\right) = \frac{\varphi(x_1, x_2)}{x_2^4},$$

where

$$\varphi(x_1, x_2) = a_0 x_1^4 + 4a_1 x_1^3 x_2 + 6a_2 x_1^2 x_2^2 + 4a_3 x_1 x_2^3 + a_4 x_2^4.$$

In this case we must consider instead of (2) a transformation of one pair of variables into another:

$$\begin{aligned} x_1 &= \alpha y_1 + \beta y_2 \\ x_2 &= \gamma y_1 + \delta y_2, \end{aligned} \qquad \left(D = \begin{vmatrix} \alpha & \beta \\ \gamma & \delta \end{vmatrix} \neq 0\right). \tag{4}$$

Here we arrive at the equality

$$\varphi(x_1, x_2) = b_0 y_1^4 + 4b_1 y_1^3 y_2 + 6b_2 y_1^2 y_2^2 + 4b_3 y_1 y_2^3 + b_4 y_2^4 = \psi(y_1, y_2),$$

and to obtain the polynomial $\psi(y)$ it remains to let

$$y = \frac{y_1}{y_2}, \qquad \psi(y) = \psi\left(\frac{y_1}{y_2}, 1\right) = \frac{\psi(y_1, y_2)}{y_2^4}.$$

The following relations hold in this notation

$$b_0 b_4 - 4b_1 b_3 + 3b_2^2 = D^4(a_0 a_4 - 4a_1 a_3 + 3a_2^2) \tag{5}$$

and

$$\begin{vmatrix} b_0 & b_1 & b_2 \\ b_1 & b_2 & b_3 \\ b_2 & b_3 & b_4 \end{vmatrix} = D^6 \begin{vmatrix} a_0 & a_1 & a_2 \\ a_1 & a_2 & a_3 \\ a_2 & a_3 & a_4 \end{vmatrix}. \tag{6}$$

For a proof we note that the general transformation (4) can be obtained by taking a composition of (at most four) partial transformations of the form

$$\text{(I)}\quad \begin{aligned} x_1 &= Dy_1, \\ x_2 &= y_2; \end{aligned} \qquad \text{(II)}\quad \begin{aligned} x_1 &= -y_2, \\ x_2 &= y_1; \end{aligned} \qquad \text{(III)}\quad \begin{aligned} x_1 &= y_1 + Ay_2, \\ x_2 &= y_2. \end{aligned}$$

Since the determinant of a general transformation is equal to the product of the determinants of the simple transformations from which it is formed, it suffices to see that (5) and (6) hold for each of these simpler transformations. Our assertion is almost trivial for the transformations (I) and (II). In the case of the transformation (III) we have the equalities

$$b_0 = a_0, \qquad b_1 = a_0 A + a_1, \qquad b_2 = a_0 A^2 + 2a_1 A + a_2,$$
$$b_3 = a_0 A^3 + 3a_1 A^2 + 3a_2 A + a_3,$$
$$b_4 = a_0 A^4 + 4a_1 A^3 + 6a_2 A^2 + 4a_3 A + a_4,$$

with the help of which it is very simple to verify the relations (5) and (6) (with $D = 1$).

A function $\Phi(a_0, a_1, a_2, a_3, a_4)$ of the coefficients of the form $\varphi(x_1, x_2)$ is called a *relative invariant* of weight m if for all transformations of the form (4)

$$\Phi(b_0, b_1, b_2, b_3, b_4) = D^m \Phi(a_0, a_1, a_2, a_3, a_4).$$

If the weight is equal to zero, then the invariant is said to be *absolute*.

In view of (5) and (6), the expressions g_2 and g_3 defined by

$$g_2 = a_0 a_4 - 4a_1 a_3 + 3a_2^2 \tag{7}$$

and

$$g_3 = \begin{vmatrix} a_0 & a_1 & a_2 \\ a_1 & a_2 & a_3 \\ a_2 & a_3 & a_4 \end{vmatrix}, \tag{8}$$

are relative invariants of the form $\varphi(x_1, x_2)$ of weight 4 and weight 6, respectively.

The quantity

$$J = \frac{g_2^3}{g_2^3 - 27g_3^2}$$

is an absolute invariant.

Let us now return from forms to polynomials.

We prove the following proposition: *for every polynomial* (1) *there is a linear fractional transformation*

$$x = \frac{\alpha y + \beta}{\gamma y + \delta}$$

with determinant

$$\begin{vmatrix} \alpha & \beta \\ \gamma & \delta \end{vmatrix} = 1$$

such that

$$\varphi(x) = \frac{4y^3 - g_2 y - g_3}{(\gamma y + \delta)^4},$$

where g_2 and g_3 are defined by (7) and (8).

Taking some transformation (2), we find that

$$\varphi(x) = \frac{b_0 y^4 + 4b_1 y^3 + 6b_2 y^2 + 4b_3 y + b_4}{(\gamma y + \delta)^4}.$$

It follows from the considerations at the beginning of the section that the transformation (2) can be chosen so that $b_0 = b_2 = 0$ and $b_1 = 1$. Using (5) and (6), we find that

$$-4b_3 = D^4 g_2, \qquad -b_4 = D^6 g_3,$$

where D is the determinant of the transformation (2). Thus,

$$\varphi(x) = \frac{4y^3 - D^4 g_2 y - D^6 g_3}{(\gamma y + \delta)^4}.$$

If $D = 1$, then out proposition is proved. If $D \neq 1$, we pass from y to a new variable (call it y_1) by the formula $y = D^2 y_1$, so that

$$x = \frac{\alpha y + \beta}{\gamma y + \delta} = \frac{\alpha D^2 y_1 + \beta}{\gamma D^2 y_1 + \delta} = \frac{\alpha_1 y_1 + \beta_1}{\gamma_1 y_1 + \delta_1},$$

where we have set

$$\alpha_1 = D^{1/2}\alpha, \qquad \beta_1 = D^{-3/2}\beta,$$
$$\gamma_1 = D^{1/2}\gamma, \qquad \delta_1 = D^{-3/2}\delta.$$

Now

$$\begin{vmatrix} \alpha_1 & \beta_1 \\ \gamma_1 & \delta_1 \end{vmatrix} = 1,$$

and, on the other hand,

$$\varphi(x) = \frac{4y_1^3 - g_2 y_1 - g_3}{(\gamma_1 y_1 + \delta_1)^4}.$$

§8. Modular forms

As we know, two numbers 2ω and $2\omega'$ whose ratio

$$\tau = \omega'/\omega \tag{1}$$

has nonzero imaginary part give rise to a regular system of points on the plane. The same regular system of points can be obtained by starting with certain other pairs of numbers.

It is not hard to see that pairs $(2\omega, 2\omega')$ and $(2\widetilde{\omega}, 2\widetilde{\omega}')$ give rise to the same regular system if the numbers 2ω and $2\omega'$ are linear combinations with integer coefficients of the numbers $2\widetilde{\omega}$ and $2\widetilde{\omega}'$, and the numbers $2\widetilde{\omega}$ and $2\widetilde{\omega}'$ are analogous combinations of the numbers 2ω and $2\omega'$. For this to hold it is necessary and sufficient that

$$\widetilde{\omega}' = \alpha\omega' + \beta\omega, \qquad \widetilde{\omega} = \gamma\omega' + \delta\omega, \tag{2}$$

where α, β, γ, δ are integers connected by the relation

$$\alpha\delta - \beta\gamma = \pm 1. \tag{3}$$

If we require that the imaginary part of the ration $\widetilde{\omega}'/\widetilde{\omega}$ has the same sign as the imaginary part of ω'/ω, then the minus sign has to be rejected in (3)[7] In this case the pairs $(2\omega, 2\omega')$ and $(2\widetilde{\omega}, 2\widetilde{\omega}')$ are said to be *equivalent*.

The quantities

$$g_2 = 60 \sum_{m,m'}' \frac{1}{(2m\omega + 2m'\omega')^4} = g_2(\omega, \omega'),$$

$$g_3 = 140 \sum_{m,m'}' \frac{1}{(2m\omega + 2m'\omega')^6} = g_3(\omega, \omega'),$$

to which we came in §6 by starting from a pair of primitive periods are relative invariants (see §7) of the polynomial in $\wp(u)$ that represents $[\wp'(u)]^2$.

We now regard these quantities as functions of the pair $(2\omega, 2\omega')$. As is easily seen, they do not change if instead of the pair $(2\omega, 2\omega')$ we take another pair generating the same regular system of points. In particular, g_2 and g_3 do not change upon passing from the pair $(2\omega, 2\omega')$ to an equivalent pair $(2\widetilde{\omega}, 2\widetilde{\omega}')$. On the other hand, it follows immediately from the definition of g_2 and g_3 that

$$g_2(t\omega, t\omega') = g_2(\omega, \omega')/t^4,$$
$$g_3(t\omega, t\omega') = g_3(\omega, \omega')/t^6.$$

Replacement of the pair $(2\omega, 2\omega')$ by $(2t\omega, 2t\omega')$ corresponds to the transition from the original network to a similar network. As we see, g_2 and

---

[7]Indeed, if $\widetilde{\tau} = (\alpha\tau + \beta)/(\gamma\tau + \delta)$, then

$$\Im\widetilde{\tau} = \Im\tau \frac{\alpha\delta - \beta\gamma}{|\gamma\tau + \delta|^2}.$$

g_3 are not invariant under such transformations. However, the quantity

$$J = \frac{g_2^3}{g_2^3 - 27g_3^2}$$

clearly remains unchanged not only under the transition from the pair $(2\omega, 2\omega')$ to an equivalent pair $(2\widetilde{\omega}, 2\widetilde{\omega}')$, but also under the transition from $(2\omega, 2\omega')$ to the pair $(2t\omega, 2t\omega')$, which generates a similar network. This quantity J, called an absolute invariant in §7, is thus a function of a single variable, namely, of the ratio $\tau = \omega'/\omega$, and it has the following property: *for any integers α, β, γ, and δ such that*

$$\alpha\delta - \beta\gamma = 1, \tag{4}$$

we have the equality

$$J\left(\frac{\alpha\tau + \beta}{\gamma\tau + \delta}\right) = J(\tau).$$

The linear substitution

$$\tau' = \frac{\alpha\tau + \beta}{\gamma\tau + \delta},$$

where α, β, γ are integers connected by the relation (4),[8] is called a *modular substitution.*

An analytic function that is invariant under modular substitutions is called a *modular function.* It will be proved below that $J(\tau)$ is an analytic function. Therefore, $J(\tau)$ is a modular function. As for the invariants g_2 and g_3, which are not functions of τ, it is natural to call them *modular forms* of ω and ω'.

Modular substitutions will be denoted by the letters $S, T, \ldots$. For example, if

$$\tau' = \frac{\alpha\tau + \beta}{\gamma\tau + \delta},$$

then we write

$$\tau' = S\tau \tag{5}$$

and

$$S = \begin{pmatrix} \alpha & \beta \\ \gamma & \delta \end{pmatrix} = \begin{pmatrix} -\alpha & -\beta \\ -\gamma & -\delta \end{pmatrix}.$$

Thus, here we do not distinguish between the two matrices[9]

$$\begin{pmatrix} \alpha & \beta \\ \gamma & \delta \end{pmatrix}, \qquad \begin{pmatrix} -\alpha & -\beta \\ -\gamma & -\delta \end{pmatrix}$$

---

[8] The relation (4) expresses that the determinant of this substitution is equal to 1.

[9] This is permissible, because the only operation that will be performed on matrices is multiplication.

The identity substitution $\tau' = \tau$ is also modular. It is denoted by the letter I:

$$I = \begin{pmatrix} 1 & 0 \\ 0 & 1 \end{pmatrix} = \begin{pmatrix} -1 & 0 \\ 0 & -1 \end{pmatrix}.$$

The expression (5) underscores that τ' is regarded as the result of applying some operation to τ.

If

$$\tau' = \frac{\alpha\tau + \beta}{\gamma\tau + \delta} = S\tau,$$

then

$$\tau = \frac{-\delta\tau' + \beta}{\gamma\tau' - \alpha} = \frac{\delta\tau' - \beta}{-\gamma\tau' + \alpha}.$$

The substitution

$$\begin{pmatrix} -\delta & \beta \\ \gamma & -\alpha \end{pmatrix} = \begin{pmatrix} \delta & -\beta \\ -\gamma & \alpha \end{pmatrix},$$

which also is modular, is called the *inverse* of the substitution

$$\begin{pmatrix} \alpha & \beta \\ \gamma & \delta \end{pmatrix} = S$$

and denoted by S^{-1}.

Applying to τ a modular substitution S_1, and then applying to the result, i.e., to $S_1\tau$, a modular substitution S_2, we get some τ'. It is easy to express τ' in terms of τ. Let $\tau^* = S_1\tau$ and $\tau' = S_2\tau^*$. Then

$$\begin{aligned}
\tau' &= \frac{\alpha_2\tau^* + \beta_2}{\gamma_2\tau^* + \delta_2} = \frac{\alpha_2(\alpha_1\tau + \beta_1) + \beta_2(\gamma_1\tau + \delta_1)}{\gamma_2(\alpha_1\tau + \beta_1) + \delta_2(\gamma_1\tau + \delta_1)} \\
&= \frac{(\alpha_2\alpha_1 + \beta_2\gamma_1)\tau + (\alpha_2\beta_1 + \beta_2\delta_1)}{(\gamma_2\alpha_1 + \delta_2\gamma_1) + (\gamma_2\beta_1 + \delta_2\delta_1)}.
\end{aligned}$$

We see that τ' can be obtained by applying to τ some substitution S. The matrix of this substitution,

$$\begin{pmatrix} \alpha_2\alpha_1 + \beta_2\gamma_1 & \alpha_2\beta_1 + \beta_2\delta_1 \\ \gamma_2\alpha_1 + \delta_2\gamma_1 & \gamma_2\beta_1 + \delta_2\delta_1 \end{pmatrix},$$

is the product of the matrices of the substitutions S_2 and S_1. Therefore, the determinant of the matrix of S is equal to 1, i.e., S is also a modular substitution. This substitution S, the result of composing S_2 and S_1, is commonly called the *product* of S_2 and S_1. Here we write

$$S = S_2 S_1 \quad \text{and} \quad \tau' = (S_2 S_1)\tau = S_2 S_1 \tau$$

if $\tau' = S_2(S_1\tau)$.

Multiplying the substitution in the other order, we get that

$$\begin{pmatrix} \alpha_1 & \beta_1 \\ \gamma_1 & \delta_1 \end{pmatrix} \begin{pmatrix} \alpha_2 & \beta_2 \\ \gamma_2 & \delta_2 \end{pmatrix} = \begin{pmatrix} \alpha_1\alpha_2 + \beta_1\gamma_2 & \alpha_1\beta_2 + \beta_1\delta_2 \\ \gamma_1\alpha_2 + \delta_1\gamma_2 & \gamma_1\beta_2 + \delta_1\delta_2 \end{pmatrix}.$$

We see that $S_1 S_2 \neq S_2 S_1$ in general, i.e., the multiplication operation is not commutative. Therefore, it is necessary to distinguish between multiplication by a substitution from the right and multiplication from the left.

With respect to this multiplication operation the collection of all modular substitutions forms a *group*, and the inverse element of an S is S^{-1}. Indeed,

$$SS^{-1} = \begin{pmatrix} -\alpha\delta + \beta\gamma & \alpha\beta - \beta\alpha \\ -\gamma\delta + \delta\gamma & \beta\gamma - \alpha\delta \end{pmatrix} = \begin{pmatrix} -1 & 0 \\ 0 & -1 \end{pmatrix} = I$$

and $S^{-1}S = I$. The function $J(\tau)$ is invariant under this group of transformations. It is often necessary to consider other groups of linear fractional transformations. An analytic function that is invariant under such a transformation group is always called an *automorphic function*. Thus, the absolute invariant $J(\tau)$ is an example of an automorphic function. Periodic functions are simpler examples of automorphic functions.

§9. Fundamental regions of the group Σ

It suffices to study a doubly periodic function in some period parallelogram. The group of substitutions with respect to which a doubly periodic function is invariant is generated by the two basic substitutions:

$$\left. \begin{array}{l} S: \ \tilde{u} = u + 2\omega, \\ S': \ \tilde{u} = u + 2\omega', \end{array} \right\} \tag{1}$$

i.e., every substitutions of this group is the result of composing (multiplying) these substitutions.

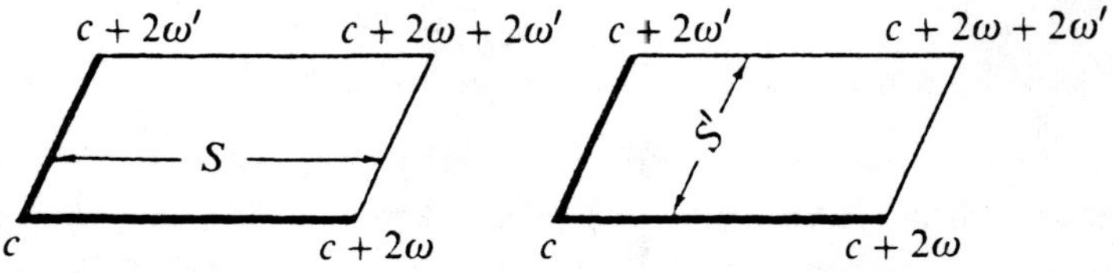

FIGURE 3

Each of the basic substitutions S and S' connects a pair of opposite sides of a period parallelogram (Figure 3). Applying to this parallelogram all the substitutions in the group, we get an infinite set of congruent parallelograms covering the whole plane once.

For each point u of the plane, a period parallelogram contains one and only one point u' that is congruent to u modulo the periods, or, in other words, *is equivalent to u with respect to the group*. Therefore, a period parallelogram is a *fundamental region* of the group under consideration.

We turn now to the modular function $J(\tau)$. The group of modular substitutions is denoted by Σ ($J(\tau)$ is invariant with respect to Σ). We show that Σ is generated by the two basic substitutions

$$S = \begin{pmatrix} 1 & 1 \\ 0 & 1 \end{pmatrix}, \qquad \tilde{\tau} = \tau + 1,$$
$$T = \begin{pmatrix} 0 & -1 \\ 1 & 0 \end{pmatrix}, \qquad \tilde{\tau} = -1/\tau. \tag{2}$$

Let $V = \begin{pmatrix} \alpha & \beta \\ \gamma & \delta \end{pmatrix}$ be an arbitrary substitution in Σ.

Using the rule for multiplying substitutions, we get that

$$VT = \begin{pmatrix} \beta & -\alpha \\ \delta & -\gamma \end{pmatrix},$$

and also that

$$VS = \begin{pmatrix} \alpha & \beta + \alpha \\ \gamma & \delta + \gamma \end{pmatrix}, \qquad VS^{-1} = \begin{pmatrix} \alpha & \beta - \alpha \\ \gamma & \delta - \gamma \end{pmatrix}$$

and in general, for an arbitrary integer n,

$$VS^{-n} = \begin{pmatrix} \alpha & \beta - n\alpha \\ \gamma & \delta - n\gamma \end{pmatrix}.$$

We successively apply two operations, namely, multiplication of a substitution (from the right) by some power of the substitution S, and multiplication by T, and we show that by starting from an (arbitrary) substitution V, we can arrive in this way at a substitution

$$VS^{-n}TS^{-m}T \cdots TS^{-k} = \begin{pmatrix} \alpha^* & 0 \\ \gamma^* & \delta^* \end{pmatrix}$$

for which $\beta^* = 0$.

If $\beta = 0$, then the original substitution already has the required property. Assuming that $\beta \neq 0$, we determine an integer n such that $|\beta - n\alpha| < |\alpha|$. After n has been found, we consider the substitution

$$V_1 = VS^{-n} = \begin{pmatrix} \alpha & \beta - n\alpha \\ \gamma & \delta - n\gamma \end{pmatrix} = \begin{pmatrix} \alpha_1 & \beta_1 \\ \gamma_1 & \delta_1 \end{pmatrix};$$

here $|\beta_1| < |\alpha_1|$. If $\beta_1 = 0$, then it is the desired substitution. But if $\beta_1 \neq 0$, then we multiply the substitution V_1 by TS^{-m}, where m is an integer, and as a result we get the substitution

$$V_2 = V_1 TS^{-m} = \begin{pmatrix} \beta_1 & -\alpha_1 - m\beta_1 \\ \delta_1 & -\gamma_1 - m\delta_1 \end{pmatrix} = \begin{pmatrix} \alpha_2 & \beta_2 \\ \gamma_2 & \delta_2 \end{pmatrix},$$

where m is chosen so that $|\alpha_1 + m\beta_1| < |\beta_1|$. Thus, for V_2 we have $|\beta_2| < |\alpha_2|$. But since $|\alpha_2| = |\beta_1|$, it follows that $|\beta_2| < |\beta_1|$. Continuing these operations, we get substitutions

$$V_3 = \begin{pmatrix} \alpha_3 & \beta_3 \\ \gamma_3 & \delta_3 \end{pmatrix}, \quad V_4 = \begin{pmatrix} \alpha_4 & \beta_4 \\ \gamma_4 & \delta_4 \end{pmatrix}, \ldots$$

with $|\beta_3| > |\beta_4| > \cdots$. Since $\beta_1, \beta_2, \ldots$ are integers, we arrive after finitely many operations at a substitution of the form we need:

$$VS^{-n}TS^{-m} \cdots TS^{-k} = \begin{pmatrix} \alpha^* & 0 \\ \gamma^* & \delta^* \end{pmatrix}.$$

Since this substitution is modular, it follows that $\alpha^* = \delta^* = 1$. Consequently,

$$VS^{-n}TS^{-m} \cdots TS^{-k} = \begin{pmatrix} 1 & 0 \\ -i & 1 \end{pmatrix},$$

where i is some integer. But $\begin{pmatrix} 1 & 0 \\ -i & 1 \end{pmatrix} = TS^iT$; therefore,

$$VS^{-n}TS^{-m}T \cdots TS^{-k} = TS^iT,$$

from which, multiplying from the right by $S^kT \cdots TS^mTS^n$, we get that

$$V = TS^iTS^kT \cdots S^mTS^n.$$

Thus, it is proved that Σ is generated by the substitutions (2).

To get a fundamental region of the group Σ, we construct in the upper half-plane a triangle with sides

$$\mathfrak{R}\tau = -1/2 \qquad \mathfrak{R}\tau = 1/2, \qquad |\tau| = 1.$$

We define the region D to be the collection of all points lying inside this triangle, along with the points lying on the left side $\mathfrak{R}\tau = -1/2$, and the points lying on the circle $|\tau| = 1$ such that $-1/2 \leq \mathfrak{R}\tau \leq 0$. Thus, D can be regarded as a quadrangle (Figure 4), with only two of the four sides included (the thick lines in the figure).

The basic substitutions (2) connect the pairs of sides of the quadrangle; namely, S connects the vertical sides, and T carries the left arc of the circle into the right arc, as pictured in Figure 4.

We prove that D is a fundamental region of the group Σ.

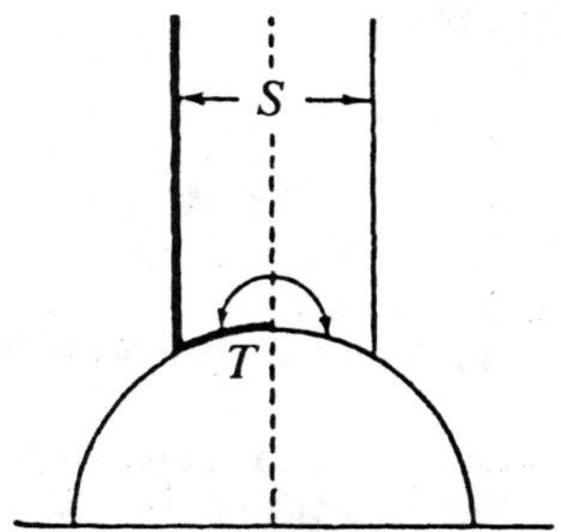

FIGURE 4

By definition, this means that for every point τ of the upper half-plane there is one and only one equivalent point τ' in D, where two points τ and τ' are said to be *equivalent* if Σ contains a substitution V such that $\tau' = V\tau$.

Let τ be a point with $\Im\tau > 0$. Take the pair $(1, \tau)$ of numbers and consider the regular system of points on the plane generated by this pair. Let all the points $m\tau + n$ in this regular system be numbered in order of nondecreasing modulus $|m\tau + n|$. We get a sequence([10])

$$0, w_1, w_2, w_3, \ldots \qquad (w_2 = -w_1). \qquad (1)$$

In this sequence we take the first point that does not lie on the line joining the origin 0 to the point w_1. Let this be the point w_k, so that

$$|w_k| \geq |w_1|. \qquad (2)$$

Both the points $w_k \pm w_1$, which occur in the sequence (1), have in it indices greater than k, because these points do not lie on the indicated line. Therefore,

$$|w_k + w_1| \geq |w_k|, \qquad |w_k - w_1| \geq |w_k|. \qquad (3)$$

We can assume that

$$\Im(w_k/w_1) > 0,$$

since $\Im(w_k/w_1) \neq 0$, and if $\Im(w_k/w_1) < 0$ we could replace w_1 by $-w_1$ (in other words, interchange the elements w_1 and w_2). As follows from its construction, the closed parallelogram with vertices $0, w_1, w_k + w_1, w_k$ does not contain points of the regular system other than its vertices. Therefore, every point of the regular system can be represented in the form $mw_1 + m'w_k$ with integers m and m'. Hence, the pair (w_1, w_k) is equivalent to the pair $(1, \tau)$.

Let $w_k = \alpha\tau + \beta$ and $w_1 = \gamma\tau + \delta$, where α, β, γ, and δ are integers. Here $\alpha\delta - \beta\gamma = +1$, since $\Im\tau > 0$ and $\Im(w_k/w_1) > 0$. We now let $w_k/w_1 = \tilde\tau$, so that

$$\tilde\tau = \frac{\alpha\tau + \beta}{\gamma\tau + \delta} = V\tau,$$

where $V \in \Sigma$.

On the basis of the inequalities (2) and (3),

$$|\tilde\tau| \geq 1, \qquad |\tilde\tau + 1| \geq |\tilde\tau|, \qquad |\tilde\tau - 1| \geq |\tilde\tau|.$$

([10])It can be assumed in general that $w_{2k} = -w_{2k-1}$.

Consequently, the point $\tilde{\tau}$ lies in the closed "triangle" with sides

$$\Re\tau = -1/2, \qquad |\tau| = 1, \qquad \Re\tau = 1/2 \qquad (\Im\tau > 0).$$

If it turns out that $\Re\tilde{\tau} \neq 1/2$, $|\tilde{\tau}| > 1$, or $-1/2 \leq \Re\tilde{\tau} \leq 0$, $|\tilde{\tau}| = 1$, then the point $\tilde{\tau}$ lies in D and thus is the desired point: $\tau' = \tilde{\tau}$. If $\Re\tilde{\tau} = 1/2$, $|\tilde{\tau}| > 1$, then $\tau' = \tilde{\tau} - 1$ is the desired point. Finally, if $0 < \Re\tilde{\tau} \leq 1/2$, $|\tilde{\tau}| = 1$, then $\tau' = -1/\tilde{\tau}$ is the desired point.

Thus, it is proved that for every point τ in the upper half-plane there is an equivalent point $\tau' \in D$.

We now prove that D does not contain equivalent points. Assume the contrary, and let τ and τ' be equivalent points in D. They cannot be connected by a transformation S^k nor by the transformation T. Hence,

$$\tau' = \frac{\alpha\tau + \beta}{\gamma\tau + \delta} \qquad \left(\neq -\frac{1}{\tau}\right),$$

and $\gamma > 0$. Since

$$\tau' - \frac{\alpha}{\gamma} = -\frac{\alpha\delta - \beta\gamma}{\gamma(\gamma\tau + \delta)},$$

it follows that

$$\tau' - \frac{\alpha}{\gamma} = -\frac{1}{\gamma(\gamma\tau + \delta)},$$

from which

$$\left|\tau' - \frac{\alpha}{\gamma}\right| \cdot \left|\tau + \frac{\delta}{\gamma}\right| = \frac{1}{\gamma^2}. \tag{4}$$

By assumption, both points τ and τ' lie in D, and the numbers $|\tau' - \alpha/\gamma|$ and $|\tau + \delta/\gamma|$ represent the distances between these points and some points on the real axis. Consequently, each of these numbers is $\geq \sqrt{3}/2$. This leads us to conclude that $\gamma = 1$, and relation (4) takes the form

$$|\tau' - \alpha| \cdot |\tau + \delta| = 1. \tag{5}$$

The distance from a point in D to an integer point of the real axis is ≥ 1. Therefore, it follows from (5) that

$$|\tau' - \alpha| = 1, \qquad |\tau + \delta| = 1.$$

From this, $\alpha = 0$ or -1, and $\delta = 0$ or 1. For $\alpha = -1$

$$\tau' = -\tfrac{1}{2} + i\sqrt{3}/2,$$

and for $\delta = 1$

$$\tau = -\tfrac{1}{2} + i\sqrt{3}/2. \tag{6}$$

Consequently, the possibility $\alpha = -1$ and $\delta = 1$ is excluded. But if $\alpha = 0$, then $\delta \neq 0$, because $|\alpha| + |\delta| \neq 0$. Therefore, for $\alpha = 0$ we must have that $\delta = 1$ and $\beta = -1$ (since $\gamma = 1$), i.e.,

$$\tau' = -\frac{1}{\tau + 1},$$

which, by (6), implies that

$$\tau' = \frac{1}{\tfrac{1}{2} + i\sqrt{3}/2} = -\tfrac{1}{2} + i\frac{\sqrt{3}}{2} = \tau.$$

Hence, this possiblity is also excluded, and it can be established similarly that the equalities $\alpha = -1$ and $\delta = 0$ are excluded.

Accordingly, our assertion is proved.

If we subject the region D to all the substitutions in Σ, then we get an infinite set of regions that are equivalent to D. They cover the whole upper half-plane, because, by what was proved, an arbitrary point of the upper half-plane has an equivalent point in D. Moreover, these regions do not overlap, since otherwise there would be at least two points τ_1 and τ_2 in the upper half-plane that could each be carried into D by two different substitutions V' and V'' in Σ. But since both equalities $V'\tau_1 = V''\tau_1$ and $V'\tau_2 = V''\tau_2$ are impossible, because the roots of the quadratic equation $V'\tau = V''\tau$ are conjugate, it follows that we would obtain two equivalent but distinct points in D ($V'\tau_1, V''\tau_1$ or $V'\tau_2, V''\tau_2$), and this is impossible.

It follows from what was proved that each region into which D is transformed by a function in Σ is also a fundamental region of Σ. We note that a fundamental region of the group Σ is often called a fundamental region of the function $J(\tau)$.

§10. The modular function $J(\tau)$

We show that $J(\tau)$ is regular at each point of the upper half-plane. Since

$$J(\tau) = \frac{g_2^3}{g_2^3 - 27g_3^2},$$

where we can assume by homogeneity that $2\omega = 1$ and $2\omega' = \tau$, and hence that

$$g_2 = 60 {\sum_{m,m'}}' \frac{1}{(m + m'\tau)^4} \equiv g_2(\tau),$$

$$g_3 = 140 {\sum_{m,m'}}' \frac{1}{(m + m'\tau)^6} \equiv g_3(\tau),$$

and since

$$g_2^3 - 27g_3^2 \neq 0$$

in the upper half-plane, it suffices to verify the regularity of the functions $g_2(\tau)$ and $g_3(\tau)$ in the upper half-plane. With this goal we verify that the series

$$ {\sum_{m,m'}}' \frac{1}{|m + m'\tau|^p}, \tag{1}$$

where $p > 2$, converges uniformly in the whole half-plane

$$\Im\tau \geq \delta > 0. \tag{2}$$

But this follows from the considerations at the beginning of §5, in view of which the sum S_n of the terms in (1) corresponding to the nth "framing" of the point $m = 0$, $m' = 0$ satisfies the inequality

$$S_n \le \frac{8}{n^{p-1}} \frac{1}{d^p},$$

where $d = \min(1, \delta)$.

We remark that the functions

$$S_{2l}(\tau) = \sideset{}{'}\sum_{m,m'} \frac{1}{(m + m'\tau)^{2l}} \qquad (\mathfrak{J}\tau > 0) \tag{3}$$

with $l \ge 2$ an arbitrary integer turn out to be useful in number theory. They are called *Eisenstein series*.

We now regard $J(\tau)$ as a function of $h^2 = e^{2\pi i\tau}$. Since $J(\tau + 1) = J(\tau)$, it follows that $J(\tau)$ is a single-valued function of h^2 ($|h| < 1$).

We prove that the expansion

$$J(\tau) = \frac{1}{h^2}(c_0 + c_1 h^2 + c_2 h^4 + \cdots)$$

is valid for $|h| < 1$, where $c_0 \ne 0$.([11])

With this goal we take the known expansion (see Table I in the back of the book)

$$\pi \cot \pi u = \frac{1}{u} + \sideset{}{'}\sum_{m} \left\{ \frac{1}{u + m} - \frac{1}{m} \right\}.$$

Let $w = e^{2\pi i u}$. Then for $|w| < 1$

$$\cot \pi u = i\frac{w + 1}{w - 1} = -i(1 + 2w + 2w^2 + \cdots),$$

and hence

$$\frac{1}{u} + \sideset{}{'}\sum_{m} \left\{ \frac{1}{u + m} - \frac{1}{m} \right\} = -\pi i - 2\pi i \sum_{k=1}^{\infty} w^k.$$

Differentiating $q \ge 2$ times with respect to u, we get

$$(-1)^q q! \sum_{m=-\infty}^{\infty} \frac{1}{(u + m)^{q+1}} = -(2\pi i)^{q+1} \sum_{k=1}^{\infty} k^q w^k$$

and set $u = n\tau$ in it ($n > 0$). This gives us that

$$(-1)^q q! \sum_{m=-\infty}^{\infty} \frac{1}{(m + n\tau)^{q+1}} = -(2\pi i)^{q+1} \sum_{k=1}^{\infty} k^q e^{2kn\pi i\tau}.$$

Summing over n from 1 to ∞, we arrive at

$$(-1)^q q! \sum_{n=1}^{\infty} \sum_{m=-\infty}^{\infty} \frac{1}{(m + n\tau)^{q+1}} = -(2\pi i)^{q+1} \sum_{k=1}^{\infty} k^q \frac{e^{2k\pi i\tau}}{1 - e^{2k\pi i\tau}}.$$

([11])We obtain in passing that $c_0 = 1: 1728 = 1: 12^3$.

We now assume that the number q is odd: $q = 2l - 1$. In this case we can rewrite the equality obtained in the form[12]

$$\frac{1}{2}\sum_{m,n}{}' \frac{1}{(m+n\tau)^{2l}} - \sum_{m=1}^{\infty}\frac{1}{m^{2l}} = \frac{(2\pi i)^{2l}}{(2l-1)!}\sum_{k=1}^{\infty} k^{2l-1}\frac{e^{2k\pi i\tau}}{1 - e^{2k\pi i\tau}}. \tag{4}$$

Our formula (4) is needed only for $l = 2$ and $l = 3$. Taking these values l, we find that

$$g_1(\tau) = 120\left\{\sum_{m=1}^{\infty}\frac{1}{m^4} + \frac{(2\pi)^4}{3!}\sum_{k=1}^{\infty}k^3\frac{e^{2k\pi i\tau}}{1 - e^{2k\pi i\tau}}\right\},$$

$$g_3(\tau) = 280\left\{\sum_{m=1}^{\infty}\frac{1}{m^6} - \frac{(2\pi)^6}{5!}\sum_{k=1}^{\infty}k^5\frac{e^{2k\pi i\tau}}{1 - e^{2k\pi i\tau}}\right\}. \tag{5}$$

The sums $\sum_{m=1}^{\infty} 1/m^{2l}$ can be expressed in terms of the so-called Bernoulli numbers. It suffices for us to know that

$$\sum_{m=1}^{\infty}\frac{1}{m^4} = \frac{(2\pi)^4}{60\cdot 4!}, \qquad \sum_{m=1}^{\infty}\frac{1}{m^6} = \frac{(2\pi)^6}{84\cdot 6!}.$$

Expanding the right-hand sides of (5) in series in powers of h^2, we get that

$$g_2 = \frac{(2\pi)^4}{2\cdot 3!}\{1 + 240h^2 + \cdots\}, \qquad g_3 = \frac{(2\pi)^6}{9\cdot 4!}\{1 - 504h^2 + \cdots\}.$$

From this,

$$g_2^3 = 27g_3^2 = \frac{(2\pi)^{12}}{8\cdot(3!)^3}\{(1 + 240h^2 + \cdots)^3 - (1 - 504h^2 - \cdots)^2\}$$

$$= \frac{(2\pi)^{12}}{8\cdot(3!)^3}\{1728h^2 + \cdots\},$$

and hence

$$J(\tau) = -\frac{g_2^3}{g_2^3 - 27g_3^2} = \frac{1 + 240h^2 + \cdots}{1728h^2 + \cdots}$$

$$= \frac{1}{1728}\frac{1}{h^2} + c_1 + c_2 h^2 + \cdots.$$

Our assertion is proved.

---

[12]Formulas (3) and (4) imply the equality

$$\frac{(2l-1)!}{(2\pi i)^{2l}}\left\{\frac{1}{2}S_{2l}(\tau) - \sum_{m=1}^{\infty}\frac{1}{m^{2l}}\right\} = \sum_{n=1}^{\infty}n^{2l-1}\frac{x^n}{1 - x^n} \tag{4'}$$

At the same time, for an arbitrary integer $q \geq 0$ and an arbitrary $|x| < 1$ we can write

$$\sum_{n=1}^{\infty}n^q\frac{x^n}{1 - x^n} = \sum_{n=1}^{\infty}\sigma_q(n)x^n.$$

It turns out that the coefficient $\sigma_q(n)$ of this series is equal to the sum of the qth powers of the positive divisors of the positive integer n. Thus, the right-hand side of (4') represents the so-called generating function for $\sigma_q(n)$ when $q = 2l - 1$. Due to (4'), certain facts in the theory of modular functions find application in the investigation of the number-theoretic function $\sigma_q(n)$.

THEOREM *For any finite c the equation*

$$J(\tau) - c = 0 \tag{6}$$

has one and only one root in the region D.

The proof is based on the fact that the number of roots of the equation $f(z) - c = 0$ in the region G bounded by a contour L is equal to

$$N = \frac{1}{2\pi i} \int_L \frac{f'(z)}{f(z) - c}\, dz$$

if $f(z)$ is regular in G, is continuous up to the boundary L, and does not take the value c on L.

Let $\tau = \xi + i\eta$. Since, by what was proved above,

$$J(\tau) = \frac{1}{1728} e^{-2\pi i\tau} + c_1 + c_2 e^{2\pi i\tau} + \cdots,$$

the function $J(\tau)$ tends to infinity uniformly with respect to ξ as $\eta \to \infty$. Consequently, for any finite c there is an H such that $|J(\tau)| > |c|$ for $\eta \geq H$, and hence equation (6) does not have roots for $\eta \geq H$. Thus, it suffices to consider the cut region D_H (bounded by the curve $MABA'M'$ (Figure 5)).

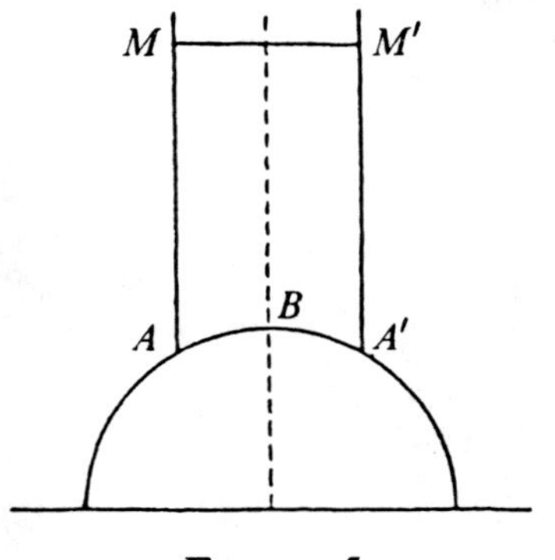

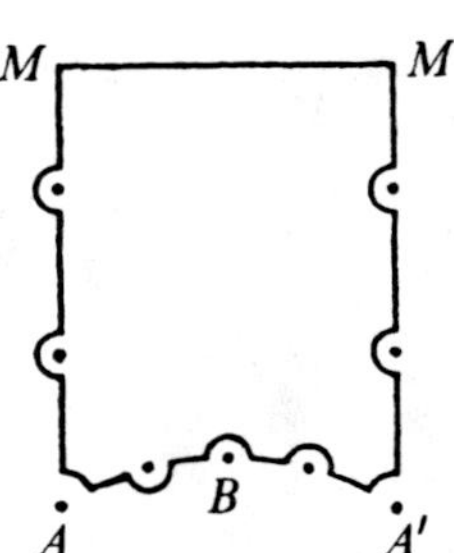

FIGURE 5   **FIGURE 6**

If (6) does not have roots on the curve $MABA'M'$, then the proof of the theorem is very simple. Indeed,

$$N = \frac{1}{2\pi i} \oint \frac{J'(\tau)}{J(\tau) - c}\, d\tau$$

$$= \frac{1}{2\pi i} \int_{MA} + \frac{1}{2\pi i} \int_{AB} + \frac{1}{2\pi i} \int_{BA'} + \frac{1}{2\pi i} \int_{A'M'} + \frac{1}{2\pi i} \int_{M'M} d\ln\{J(\tau) - c\}$$

$$= J_1 + J_2 + J_3 + J_4 + J_5.$$

The function $\varphi(\tau) = J(\tau) - c$ satisfies the relations

$$\varphi(-1/\tau) = \varphi(\tau), \qquad \varphi(\tau + 1) = \varphi(\tau). \tag{α}$$

Setting $\tau = -1/t$ in the integral J_2, we get that

$$J_2 = \frac{1}{2\pi i} \int_{AB} d\ln\varphi(\tau) = \frac{1}{2\pi i} \int_{A'B} d\ln\varphi(\tau) = -J_3.$$

Thus, $J_2 + J_3 = 0$. Analogously (and perhaps more simply), it can be proved that $J_1 + J_4 = 0$. Consequently,

$$N = \frac{1}{2\pi i} \int_{M'M} d \ln \varphi(\tau).$$

We can regard $J(\tau)$ as a function of $z = h^2 = e^{2\pi i \tau}$. The segment $M'M$ in the τ-plane corresponds in the z-plane to the circle

$$|z| = e^{-2\pi H}. \tag{7}$$

On this circle and inside it (i.e., for $\eta \geq H$) the function $\varphi(\tau)$ is nonzero and, moreover, regular if the first-order pole at the point $z = 0$ is excluded. Since integration along $M'M$ in the τ-plane reduces to integration along the circle (7) in the negative direction, it follows that

$$N = -\frac{1}{2\pi i} \int_K d \ln \varphi(\tau),$$

where K is the circle (7), traversed in the positive direction, and hence N is equal to the number of poles of the function $\varphi(\tau) = J(\tau) - c = \psi(z)$ in the disk $|z| < e^{-2\pi H}$, i.e., $N = 1$.

We now study the case when equation (6) has roots on the curve $MABA'M'$. In every case there are finitely many of these roots, and together with each of them there will be an equivalent one, namely, the number symmetric with respect to the imaginary axis. About each of these roots and about each of the points A, B, and A' we draw a circle of radius ε small enough that these circles are disjoint. With the help of these circles we modify the boundary of D_H as indicated in Figure 6, after which one of each pair of equivalent roots will line inside the contour, while the points A, B, and A' are outside the contour

Breaking up the integral N along the resulting contour into parts and using the relations (α), we prove without difficulty that

$$N = \frac{1}{2\pi i} \int_{M'M} + \frac{1}{2\pi i} \int_{\lambda_A} + \frac{1}{2\pi i} \int_{\lambda_B} + \frac{1}{2\pi i} \int_{\lambda_{A'}} d \ln\{J(\tau) - c\}.$$

Here λ_A, λ_B, and $\lambda_{A'}$ represent arcs with centers at the vertices A, B, and A'. If equation (6) does not have roots at these points, then the arcs λ_A, λ_B, $\Lambda_{A'}$ can be shrunk to points, with the result that

$$N = \frac{1}{2\pi i} \int_{M'M} d \ln\{J(\tau) - c\}.$$

This integral, as shown above, is equal to 1. This implies that the theorem holds in this case.

It remains to investigate the case when (6) has roots at vertices. For this we determine what values $J(\tau)$ takes at the vertices.

At the point A

$$\tau = -\frac{1}{2} + i\frac{\sqrt{3}}{2} \equiv \rho,$$

and since $\rho^3 = 1$, it follows that

$$\frac{g_2(\rho)}{60} = \sum{}' \frac{1}{(m + m'\rho)^4} = \frac{1}{\rho} \sum{}' \frac{1}{(m\rho^2 + m')^4}.$$

On the other hand, in view of the relation $\rho^2 + \rho + 1 = 0$

$$\sum{}' \frac{1}{(m\rho^2 + m')^4} = \sum{}' \frac{1}{(m' - m - m\rho)^4} = \sum{}' \frac{1}{(n + n'\rho)^4}.$$

Therefore,

$$\frac{g_2(\rho)}{60} = \frac{1}{\rho} \sum{}' \frac{1}{(n + n'\rho)^4} = \frac{1}{\rho} \frac{g_2(\rho)}{60},$$

from which it follows that $g_2(\rho) = 0$. Thus, $J(\rho) = 0$, and hence $J(\tau) = 0$ at the point A' $(\tau = \rho + 1 = -1/\rho)$. It can be proved similarly that $g_3(i) = 0$, and hence $J(\tau) = 1$ at the point B.

We must consequently consider the two equations

$$J(\tau) - 1 = 0, \tag{a}$$

$$J(\tau) = 0. \tag{b}$$

In the case of the first equation we can shrink the arcs Λ_A and $\lambda_{A'}$ to points, and hence

$$N = \frac{1}{2\pi i} \int_{M'M} d\ln\{J(\tau) - 1\} + \frac{1}{2\pi i} \int_{\lambda_B} d\ln\{J(\tau) - 1\}.$$

If $\bar\lambda_B$ complements λ_B to form a full circle, then

$$-\oint d\ln\{J(\tau) - 1\} = \int_{\lambda_B} + \int_{\bar\lambda_B},$$

where the integral on the left-hand side is taken in the positive direction. Making the substitution $\tau = -1/t$ in the integral $\int_{\bar\lambda_B}$, we find that $\int_{\bar\lambda_B} = \int_{\lambda_B}$. Hence,

$$N = \frac{1}{2\pi i} \int_{M'M} d\ln\{J(\tau) - 1\} - \frac{1}{4\pi i} \oint d\ln\{J(\tau) - 1\}.$$

The first term on the right-hand side is equal to 1. Let

$$\frac{1}{2\pi i} \oint d\ln\{J(\tau) - 1\} = n,$$

so that $N = 1 - n/2$. The number n is the multiplicity of the root $\tau = i$ of equation (a), and since $n/2$ is a positive integer, it follows that $n \geq 2$. On the other hand, $N \geq 0$, from which $n \leq 2$. Consequently, $n = 2$ and $N = 0$.

We see that equation (a) has only one root: $\tau = i$. This is a double root, but the region D contains only half a neighborhood of the point i, and hence it can be assumed that D contains only the single simple root $\tau = i$, while the other root of (a) at the point $\tau = i$ belong to the region D' having the arc ABA' in common with D.

Equation (b) is treated similarly. Here the points $\tau = \rho$, $-\rho^2$ are roots, and each is triple root. However, D contains only the first of these points, i.e., A. Six regions meet at this point: D and five regions equivalent to it.[13] Each of these six regions contains one sixth of a full neighborhood of A. Further, in view of the definition of a fundamental region the point A belongs to only three of the regions: D and two others. The remaining three regions do not contain A, just as the point A' does not belong to D. Consequently, the three-fold root at A belongs, with equal justification, to three regions, and thus it must be assumed that equation (b) has one simple root in D.

§11. Inversion of elliptic integrals of the first kind

Starting from primitive periods

$$2\omega, \ 2\omega' \qquad (\Im(\omega'/\omega) > 0), \tag{1}$$

we constructed the Weierstrass function $\wp(u)$ in §§5 and 6 and showed that it satisfies the differential equation

$$\wp'^2 = 4\wp^3 - g_2\wp - g_3.$$

____

[13] See Figure 7 in §23.

This equation implies that

$$u = \pm \int_{\wp}^{\infty} \frac{dt}{\sqrt{4t^3 - g_2 t - g_3}}. \tag{2}$$

The reader knows from a course in integral calculus that integrals of the form $\int R(t, w)\, dt$, where R is a rational function of its arguments and w^2 is a polynomial of third or fourth degree in t without multiple roots, bear the name *elliptic integrals*. The integral (2) above is called an *elliptic integral of the first kind*. Below we shall cover elliptic integrals in detail. Here, however, it is only important for us that the Weierstrass function $\wp(u)$ constructed for the given periods (1) is one of the limits of integration of a certain elliptic integral of the first kind, regarded as a function of the value of this integral. Further, the numbers g_2 and g_3 under the square root sign were not given arbitrarily, but were defined in terms of the periods (1), and we saw that

$$g_2^3 - 27 g_3^2 \neq 0.$$

The following question arises naturally: if numbers a_2 and a_3 with $a_2^3 - 27 a_3^2 \neq 0$ are given and the integral

$$u = \pm \int_{x}^{\infty} \frac{dt}{\sqrt{4t^3 - a_2 t - a_3}} \tag{3}$$

is considered, then is its lower limit x an elliptic function of the value of the integral?

This question has a positive answer, in the following form.

In the first place it is established that there are numbers 2ω and $2\omega'$ such that $\Im(\omega'/\omega) > 0$ and

$$g_2(\omega, \omega') = a_2, \qquad g_3(\omega, \omega') = a_3. \tag{4}$$

Then the function $\wp(v)$ with periods 2ω and $2\omega'$ is constructed. Finally, the change of variable $t = \wp(v)$ is made in the integral (3). In view of the equation

$$[\wp'(v)]^2 = 4[\wp(v)]^3 - a_2 \wp(v) - a_3,$$

we get that

$$u = \pm \int_{0}^{w} dv, \tag{5}$$

where $x = \wp(w)$. It follows from (5) that $w = \pm u$. Hence, $x = \wp(u)$, which is what was asserted.

We see that everything depends on solving the following problem: given numbers a_2 and a_3 with $a_2^3 - 27 a_3^2 \neq 0$, find 2ω and $2\omega'$ such that (4) holds.

The solution of this problem can be obtained immediately on the basis of §10. Take the equation

$$J(\tau) = \frac{a_2^3}{a_2^3 - 27a_3^2}.$$

It has a solution τ in a fundamental region, hence in the upper half-plane. Finding this solution, we determine ω from the equation

$$\frac{1}{(2\omega)^4} g_2(\tau) = a_2$$

if $a_2 \neq 0$, and from the equation

$$\frac{1}{(2\omega)^6} g_3(\tau) = a_3$$

if $a_2 = 0$. When ω has been found, ω' is determined from the equation $\omega'/\omega = \tau$.

CHAPTER 3

Weierstrass Functions

§12. The Weierstrass function $\zeta(u)$

This function is defined by the formula

$$\zeta(u) = \frac{1}{u} - \int_0^u \left\{ \wp(u) - \frac{1}{u^2} \right\} du, \tag{1}$$

so that

$$\zeta'(u) = -\wp(u). \tag{2}$$

Here the path of integration in (1) must not go through vertices of the period network other than the point $u = 0$.

Replacing $\wp(u)$ in (1) by the expansion of this function in partial fractions, we get for $\zeta(u)$ the representation

$$\zeta(u) = \frac{1}{u} + \sideset{}{'}\sum_{m,m'} \left\{ \frac{1}{u - 2m\omega - 2m'\omega'} \right.$$
$$\left. + \frac{1}{2m\omega + 2m'\omega'} + \frac{u}{(2m\omega + 2m'\omega')^2} \right\}, \tag{3}$$

which shows that the only singularities of $\zeta(u)$ are simple poles at the points $2m\omega + 2m'\omega'$. It follows from (1) that $\zeta(u)$ is an odd function. Indeed,

$$\zeta(-u) = -\frac{1}{u} - \int_0^{-u} \left\{ \wp(v) - \frac{1}{v^2} \right\} dv$$
$$= -\frac{1}{u} + \int_0^u \left\{ \wp(v) - \frac{1}{v^2} \right\} dv = -\zeta(u).$$

On the other hand, by (2),

$$\left. \begin{aligned} \zeta(u + 2\omega) &= \zeta(u) + 2\eta, \\ \zeta(u + 2\omega') &= \zeta(u) + 2\eta', \end{aligned} \right\} \tag{4}$$

35

where η and η' are constants. Setting $u = -\omega$ and $u = -\omega'$ in these equalities, respectively, and using the oddness of $\zeta(u)$, we get that $\eta = \zeta(\omega)$ and $\eta' = \zeta(\omega')$.

The notation $\eta = \eta_1$ and $\eta' = \eta_3$ is often used, so that (see the last page of Chapter I)

$$\eta_1 = \zeta(\omega_1), \qquad \eta_3 = \zeta(\omega_3),$$

and the additional constant

$$\eta_2 = \zeta(\omega_2) = -\zeta(\omega + \omega')$$

is introduced. It is not hard to see that $\eta_1 + \eta_2 + \eta_3 = 0$. Indeed, by (4),

$$\zeta(u + 2\omega + 2\omega') = \zeta(u + 2\omega') + 2\eta = \zeta(u) + 2\eta + 2\eta.$$

From this, setting $u = -\omega - \omega'$, we find that $\eta + \eta' = \zeta(\omega + \omega')$, i.e., $\eta_1 + \eta_3 = -\eta_2$.

We now prove the very important relation

$$\eta\omega' - \eta'\omega = \pi i/2, \tag{5}$$

which is valid under the condition $\mathfrak{I}(\omega'/\omega) > 0$. To get the relation (5) we take some period parallelogram for which the point $u = 0$ is an interior point. Let the points c, $c + 2\omega$, $c + 2\omega + 2\omega'$, and $c + 2\omega'$ be the vertices of the parallelogram. Integrating $\zeta(u)$ along the contour of this parallelogram, we get that

$$2\pi i = \int_c^{c+2\omega} + \int_{c+2\omega}^{c+2\omega+2\omega'} + \int_{c+2\omega+2\omega'}^{c+2\omega'} + \int_{c+2\omega'}^{c} \zeta(u)\,du.$$

Making the substitution $u = 2\omega + v$ in the second integral, and $u = 2\omega' + v$ in the third, we have that

$$2\pi i = \int_c^{c+2\omega} \{\zeta(u) - \zeta(u + 2\omega')\}\,du$$

$$+ \int_c^{c+2\omega'} \{\zeta(u + 2\omega) - \zeta(u)\}\,du$$

$$= \int_c^{c+2\omega'} 2\eta\,du - \int_c^{c+2\omega} 2\eta'\,du = 4(\eta\omega' - \eta'\omega).$$

Thus, relation (5) is proved. We shall have reason to return to it below. It can be rewritten in the form

$$\eta_1\omega_3 - \eta_3\omega_1 = \pi i/2.$$

Cyclic permutation can be used to obtain two more relations from this one:

$$\eta_2\omega_1 - \eta_1\omega_2 = \pi i/2, \qquad \eta_3\omega_2 - \eta_2\omega_3 = \pi i/2.$$

§13. The Weierstrass function $\sigma(u)$

We define the function $\sigma(u)$ with the help of the equality

$$\ln \frac{\sigma(u)}{u} = \int_0^u \left\{ \zeta(u) - \frac{1}{u} \right\} \, du, \tag{1}$$

where the path of integration misses all the vertices of the period network other than the point $u = 0$. It follows from (1) that

$$\sigma'(u)/\sigma(u) = \zeta(u). \tag{2}$$

Replacing $\zeta(u)$ in (1) by its expansion in partial fractions and integrating termwise, we get that

$$\ln \frac{\sigma(u)}{u} = {\sum_{m,m'}}' \left\{ \ln \left(1 - \frac{u}{2m\omega + 2m'\omega'} \right) \right.$$
$$\left. + \frac{u}{2m\omega + 2m'\omega'} + \frac{u^2}{2(2m\omega + 2m'\omega')^2} \right\}$$

This implies the following expansion of $\sigma(u)$ in an infinite product:

$$\sigma(u) = u {\prod}' \left(1 - \frac{u}{s} \right) e^{u/s + u^2/2s^2} \qquad (s = 2m\omega + 2m'\omega'). \tag{3}$$

We see that $\sigma(u)$ is an entire transcendental function having only simple zeros lying at the vertices of the period network.

It follows immediately from (1) or (3) that $\sigma(u)$ is an odd function.

Let us replace u by $u + 2\omega$ in (2). On the basis of (4) in §12,

$$\frac{\sigma'(u + 2\omega')}{\sigma(u + 2\omega)} = \frac{\sigma'(u)}{\sigma(u)} + 2\eta.$$

From this,

$$\ln \sigma(u + 2\omega) = \ln \sigma(u) + 2\eta u + C,$$

and hence

$$\sigma(u + 2\omega) = C' e^{2\eta u} \sigma(u).$$

Setting $u = -\omega$ here, we have that

$$\sigma(\omega) = -\sigma(\omega) C' e^{-2\eta\omega}.$$

But since $\sigma(\omega) \neq 0$, it follows that $C' = -e^{2\eta\omega}$. Hence,

$$\sigma(u + 2\omega) = -e^{2\eta(u+\omega)} \sigma(u).$$

It is easy to see that, in general,

$$\sigma(u + 2\omega_\alpha) = -e^{2\eta_\alpha(u+\omega_\alpha)} \sigma(u) \qquad (\alpha = 1, 2, 3). \tag{4}$$

§14. Expression of an arbitrary elliptic function by means of $\sigma(u)$ and by means of $\zeta(u)$

Every rational function $R(z)$ admits the following two representations:

$$R(z) = C\frac{(z - b_1)(z - b_2)\cdots(z - b_n)}{(z - a_1)(z - a_2)\cdots(z - a_m)}, \tag{α}$$

$$R(z) = E(z) + \sum_{i,k} \frac{A_k^{(i)}}{(z - a_k)^i}, \tag{β}$$

where C, b_i, a_k and $A_k^{(i)}$ are constants, and $E(z)$ is a polynomial, the so-called *entire part* of $R(z)$. Each of these representations gives definite information about $R(z)$: it is clear from the first representation what the zeros of $R(z)$ are and what the poles are, and the second representation, which is commonly used in integral calculus, gives the principal part of $R(z)$ for each of its poles.

We now show that an arbitrary elliptic function admits an analogous representation.

Let $f(u)$ be an elliptic function with periods 2ω and $2\omega'$. We take some period parallelogram and suppose that $f(u)$ has poles $a_1, \ldots, a_n$ and zeros $b_1, \ldots, b_n$ in this parallelogram. Furthermore, each zero and each pole will be written as many times as its multiplicity. As we know (see §4),

$$a_1 + \cdots + a_n \equiv b_1 + \cdots + b_n \pmod{(2\omega, 2\omega')}.$$

Let

$$b_1 = b_1^* + 2m\omega + 2m'\omega',$$

where the integers m and m' are chosen so that

$$a_1 + \cdots + a_n = b_1^* + b_2 + \cdots + b_n. \tag{1}$$

The point b_1^* does not in general lie in the period parallelogram under consideration. However, it does not do us any harm to take instead of the system of zeros $b_1, \ldots, b_n$ the equivalent system $b_1^*, b_2, \ldots, b_n$. We now construct the function

$$g(u) = \frac{\sigma(u - b_1^*)\sigma(u - b_2)\cdots\sigma(u - b_n)}{\sigma(u - a_1)\sigma(u - a_2)\cdots\sigma(u - a_n)},$$

This function has the same zeros and the same poles (also of the same multiplicity) as $f(u)$. On the other hand, by the properties of $\sigma(u)$ and relation (1),

$$g(u + 2\omega_\alpha) = e^{2\eta_\alpha(a_1 + \cdots + a_n - b_1^* - b_2 - \cdots - b_n)} g(u) = g(u),$$

so that $g(u)$ is an elliptic function with the same periods as $f(u)$. The ratio

$$f(u)/g(u) \tag{2}$$

does not have poles, because each pole of the numerator is a pole of the same multiplicity for the denominator, and each zero of the denominator is a zero of the same multiplicity for the numerator. But (2) is an elliptic function. Consequently, this ratio is a constant, and we get the first representation of $f(u)$:

$$f(u) = C\frac{\sigma(u - b_1^*)\sigma(u - b_2)\cdots\sigma(u - b_n)}{\sigma(u - a_1)\sigma(u - a_2)\cdots\sigma(u - a_n)},$$

where

$$b_1^* + b_2 + \cdots + b_n = a_1 + \cdots + a_n.$$

This representation is an analogue of (α).

We proceed to the second representation of an elliptic function. Suppose that the poles ([14]) $a_1, \ldots, a_n$ of $f(u)$ lying in some fundamental parallelogram are known, along with the corresponding principal parts of $f(u)$. Let the principal part corresponding to the pole a_k have the form

$$\frac{A_k}{u - a_k} + \sum_{r=2}^{m_k}(-1)^r\frac{(r - 1)!A_k^{(r-1)}}{(u - a_k)^r}.$$

As is easy to see, the function

$$A_k\zeta(u - a_k) + \sum_{r=2}^{m_k} A_k^{(r-1)}\wp^{(r-2)}(u - a_k)$$

has the same principal part. Forming the sum of these expressions over all poles, we get the function

$$\sum_{k=1}^{n} A_k\zeta(u - a_k) + \sum_{\substack{k,r \\ (r\geq 2)}} A_k^{(r-1)}\wp^{(r-2)}(u - a_k).$$

The second term of this sum is an elliptic function. We show that the same is true for the first term. Indeed, let

$$\varphi(u) = \sum_{k=1}^{n} A_k\zeta(u - a_k).$$

Then

$$\varphi(u + 2\omega_\alpha) = \sum_{k=1}^{n} A_k 2\eta_\alpha + \varphi(u) = 2\eta_\alpha\sum_{k=1}^{n} A_k + \varphi(u).$$

([14]) Here each pole is written *once*, and not as many times as its multiplicity.

But A_k is the residue of our elliptic function with respect to the pole a_k. And since the sum of the residues with respect to all the poles in a period parallelogram is zero, it follows that

$$\varphi(u + 2\omega_\alpha) = \varphi(u),$$

and hence $\varphi(u)$ is an elliptic function.

The difference

$$f(u) - \sum_{k=1}^{n} A_k \zeta(u - a_k) - \sum_{\substack{k,r \\ (r \geq 2)}} A_k^{(r-1)} \wp^{(r-2)}(u - a_k)$$

does not have singular points in the period parallelogram under consideration, and being an elliptic function, it is thus a constant. Accordingly, we obtain the second representation

$$f(u) = C + \sum_{k=1}^{n} A_k \zeta(u - a_k) + \sum_{k,r} A_k^{(r-1)} \wp^{(r-2)}(u - a_k)$$

for $f(u)$; it is analogous to (β) and can be called the expansion of $f(u)$ in "partial fractions".

§15. The addition theorems for Weierstrass functions

We consider the function $\wp(u) - \wp(v)$, where v is a constant quantity (of course, not congruent to zero modulo the periods 2ω and $2\omega'$). This function has a second-order pole at the point $u = 0$ and simple poles at $u = v$ and $u = -v$. By using the theorem in §14, it is easy to see that we can set $a_1 = a_2 = 0$, $b_1^* = v$, and $b_2 = -v$. Thus, we get the representation

$$\wp(u) - \wp(v) = C \frac{\sigma(u - v)\sigma(u + v)}{[\sigma(u)]^2},$$

where C is a constant. To determine this constant we multiply both sides of this relation by u^2 and let $u = 0$. This gives us that $1 = -C[\sigma(v)]^2$. Consequently,

$$C = -1/[\sigma(v)]^2,$$

and hence

$$\wp(u) - \wp(v) = -\frac{\sigma(u - v)\sigma(u + v)}{[\sigma(u)]^2[\sigma(v)]^2}. \tag{1}$$

In this equality we replace v by ω_α ($\alpha = 1, 2, 3$) and recall that (see §6) $\wp(\omega_\alpha) = e_\alpha$. Since

$$\sigma(u + \omega_\alpha) = \sigma(u - \omega_\alpha + 2\omega_\alpha) = -e^{2\eta_\alpha u}\sigma(u - \omega_\alpha),$$

we get the equality

$$\wp(u) - e_\alpha = e^{2\eta_\alpha u}\left[\frac{\sigma(u - \omega_\alpha)}{\sigma(\omega_\alpha)\sigma(u)}\right]^2 \qquad (\alpha = 1, 2, 3).$$

In addition to $\sigma(u)$ we introduce three more sigma functions:

$$\sigma_\alpha(u) = -\frac{e^{\eta_\alpha u}\sigma(u - \omega_\alpha)}{\sigma(\omega_\alpha)} \qquad (\alpha = 1, 2, 3). \tag{2}$$

The minus sign is taken so that $\sigma_\alpha(0) = 1$. Thus,

$$\wp(u) - e_\alpha = [\sigma_\alpha(u)/\sigma(u)]^2 \qquad (\alpha = 1, 2, 3).$$

We see that the square root of $\wp(u) - e_\alpha$ is a single-valued function. Once and for all we take the determination of this root that behaves like $+1/u$ in a neighborhood of $u = 0$. Then

$$\sqrt{\wp(u) - e_\alpha} = \sigma_\alpha(u)/\sigma(u) \qquad (\alpha = 1, 2, 3). \tag{3}$$

The function $\wp'(u)$ can also be expressed simply in terms of sigma functions. To get this expression we take the relation

$$\wp'^2 = 4(\wp - e_1)(\wp - e_2)(\wp - e_3).$$

This relation gives us that

$$[\wp'(u)]^2 = 4\frac{[\sigma_1(u)\sigma_2(u)\sigma_3(u)]^2}{[\sigma(u)]^6}.$$

Taking the square root and noting that

$$\lim_{u \to 0} u^3 \wp'(u) = -2,$$

we find that

$$\wp'(u) = -2\frac{\sigma_1(u)\sigma_2(u)\sigma_3(u)}{[\sigma(u)]^3}. \tag{4}$$

Let us turn again to relation (1). Taking the logarithmic derivative of both sides, we get

$$\frac{\wp'(u)}{\wp(u) - \wp(v)} = \zeta(u - v) + \zeta(u + v) - 2\zeta(u). \tag{5_1}$$

This is an expansion of the left-hand side into partial fractions. It could also have been obtained directly by using the general theorem in §14.

We interchange u and v in (5_1):

$$-\frac{\wp'(v)}{\wp(u) - \wp(v)} = -\zeta(u - v) + \zeta(u + v) - 2\zeta(v). \tag{5_2}$$

Let us now add (5_1) (5_2) termwise. This gives us

$$\frac{\wp'(u) - \wp'(v)}{\wp(u) - \wp(v)} = 2\zeta(u + v) - 2\zeta(u) - 2\zeta(v).$$

From this,

$$\zeta(u + v) = \zeta(u) + \zeta(v) + \frac{1}{2}\frac{\wp'(u) - \wp'(v)}{\wp(u) - \wp(v)}. \tag{6}$$

This equality represents the zeta function of the sum of two arguments in terms of certain functions of each argument separately. One says that (6) expresses the addition theorem for the zeta function.

To get the addition theorem for $\wp$ we differentiate(6) with respect to u and with respect to v. We have

$$-\wp(u + v) = -\wp(u)$$
$$+ \frac{1}{2}\frac{\wp''(u)[\wp(u) - \wp(v)] - \wp'(u)[\wp'(u) - \wp'(v)]}{[\wp(u) - \wp(v)]^2},$$
$$-\wp(u + v) = -\wp(v)$$
$$- \frac{1}{2}\frac{\wp''(v)[\wp(u) - \wp(v)] - \wp'(v)[\wp'(u) - \wp'(v)]}{[\wp(u) - \wp(v)]^2}.$$

Let us add these equalities termwise:

$$-2\wp(u + v) = -\wp(u) - \wp(v)$$
$$+ \frac{1}{2}\frac{[\wp''(u) - \wp''(v)][\wp(u) - \wp(v)] - [\wp'(u) - \wp'(v)]^2}{[\wp(u) - \wp(v)]^2}. \tag{7}$$

In view of the differential equation of $\wp$,

$$2\wp'' = 12\wp^2 - g_2;$$

therefore,

$$\wp''(u) - \wp''(v) = 6[\wp^2(u) - \wp^2(v)],$$

and by using this identity it is not hard to reduce (7) to the form

$$\wp(u + v) + \wp(u) + \wp(v) = \frac{1}{4}\left[\frac{\wp'(u) - \wp'(v)}{\wp(u) - \wp(v)}\right]^2, \tag{8}$$

which expresses the addition theorem for the function $\wp$.

EXERCISE 1. Using the addition theorem and the differential equation for $\wp$, prove the following identity:

$$[\wp(u - v/2) - \wp(u + v/2)]^2$$
$$= \left\{\frac{1}{4}\left[\frac{\wp'(u - v/2) - \wp'(v)}{\wp(u - v/2) - \wp(v)}\right]^2 - 3\wp(v)\right\}^2$$
$$+ 2\wp'(v)\frac{\wp'(u - v/2) - \wp'(v)}{\wp(u - v/2) - \wp(v)} + g_2 - 12[\wp(v)]^2. \tag{9}$$

EXERCISE 2. Prove the identity

$$\frac{\sigma_1(u) + \sigma_2(u) + \sigma_3(u)}{\sigma(u)} = -\frac{1}{2}\frac{\wp''(u/2)}{\wp'(u/2)} \tag{10}$$

(S. V. Kovalevskaya).

It suffices to verify that both sides of (10) are odd elliptic functions with periods 4ω and $4\omega'$ that have simple poles 0, 2ω, $2\omega'$, and $2\omega + 2\omega'$ with the same corresponding residues $(3, -1, -1, -1)$ in the period parallelogram.

§16. Representation of every elliptic function in terms of the functions $\wp(u)$ and $\wp'(u)$

It was shown in §14 that every elliptic function $f(u)$ admits an expansion into partial fractions:

$$f(u) = C + \sum_{k=1}^{n} A_k \zeta(u - a_k) + \sum_{\substack{k,r \\ (r \geq 2)}} A_k^{(r-1)} \wp^{(r-2)}(u - a_k);$$

further,

$$\sum_{k=1}^{n} A_k = 0. \tag{1}$$

We now use the addition theorems for the functions ζ and $\wp$; moreover, we take into account that every derivative of $\wp$ can be expressed rationally in terms of $\wp$ and $\wp'$.

First of all, in view of the addition theorem for ζ,

$$\sum_{k=1}^{n} A_k \zeta(u - a_k) = \sum_{k=1}^{n} A_k \zeta(u) + R_1(\wp, \wp') = R_1(\wp, \wp'), \tag{2}$$

where R_1, like the $R_2, R_3, \ldots$ encountered below, denotes a rational function of its arguments. Relation (1) is used in establishing (2).

On the basis of the addition theorem for $\wp$ we get that

$$\sum_{k=1}^{n} A_k^{(1)} \wp(u - a_k) = R_2(\wp, \wp').$$

Then

$$\sum_{k=1}^{n} A_k^{(2)} \wp'(u - a_k) = R_3(\wp, \wp'),$$

and so on. By virtue of all these equalities,

$$f(u) = R(\wp, \wp').$$

Thus, every elliptic function can be rationally expressed in terms of the functions $\wp$ and $\wp'$.

This representation can be given the form

$$f(u) = R_1(\wp) + R_2(\wp)\wp', \tag{3}$$

in which there now enter rational functions of $\wp$ alone. Indeed, in view of the differential equation for $\wp$ every positive integer power of $\wp'$ can be expressed in the form $A + B\wp'$, where A and B are polynomials in $\wp$. Therefore, a rational function of $\wp$ and $\wp'$ can be represented in the form

$$R(\wp, \wp') = \frac{M_1 + N_1\wp'}{M + N\wp'},$$

where M_1, N_1, M, and N are polynomials in $\wp$. Multiplying the denominator and numerator by $M - N\wp'$, we get that

$$R(\wp, \wp') = \frac{M_2 + N_2\wp'}{M^2 - N^2\wp'^2}.$$

The denominator is now a polynomial in $\wp$ alone. Hence, $R(\wp, \wp') = R_1(\wp) + R_2(\wp)\wp'$, which is what was required to prove.

We remark further that an even elliptic function can be represented in the form $f(u) = R(\wp)$, and an odd elliptic function in the form $f(u) = R(\wp)\wp'$. To prove this we take the representation

$$f(u) = R_1(\wp) + R_2(\wp)\wp'(u)$$

and replace u by $-u$. This gives us that

$$f(-u) = R_1(\wp) - R_2(\wp)\wp'(u),$$

because $\wp(u)$ is an even function, and $\wp'(u)$ is an odd function. It now remains to add the second formula to the first if $f(u) = f(-u)$, and to subtract if $f(u) = -f(-u)$.

Many general properties of elliptic functions follow from the representation (3).

One of the most important properties is as follows: *any two elliptic functions with the same periods are connected by an algebraic relation.*

Let $f(u)$ and $g(u)$ be two such functions. Then

$$f(u) = R_1(\wp) + R_2(\wp)\wp', \tag{4_1}$$

$$g(u) = R_3(\wp) + R_4(\wp)\wp'. \tag{4_2}$$

On the other hand,

$$\wp'^2 = 4\wp^3 - g_2\wp - g_3. \tag{5}$$

We can eliminate $\wp$ and $\wp'$ from (4_1), (4_2), and (5). This leads to a relation of the form $F(f, g) = 0$, where F is a polynomial in its two arguments.

We note two special cases of this general statement.

For the first case let $g = f'$. We then obtain the following fact: *every elliptic function satisfies a first-order differential equation of the form $F(f, f') = 0$, where F denotes a polynomial in its arguments.*

For the second special case let $g(u) = f(u+v)$. We then get the relation

$$\sum_{k=0}^{N} P_k[f(u)][f(u+v)]^k = 0,$$

where the $P_k(z)$ are polynomials in z with coefficients depending on v. Interchanging u and v and comparing the result with the original expression, we find that $P_k[f(u)]$ is a symmetric polynomial in $f(u)$ and $f(v)$ with constant coefficients.

Thus, *every elliptic function $f(u)$ satisfies an equation*

$$g(f(u), f(v), f(u+v)) = 0,$$

where g is a polynomial with constant coefficients. The existence of such a relation expresses the fact that *f has an algebraic addition theorem.*

An example of such an algebraic addition theorem is given by formula (8) in §15, since the derivatives $\wp'(u)$ and $\wp'(v)$ on its right-hand side are algebraic functions of $\wp(u)$ and $\wp(v)$, respectively. Conversely, formula (6) in §15 does not represent an algebraic addition theorem, since $\wp$ and its derivative are not expressed algebraically in terms of ζ.

§17. Elliptic integrals

Back in §11 we mentioned the general form for elliptic integrals:

$$\int R(z, w)\, dz. \tag{1}$$

Here

$$w^2 = a_0 z^4 + 4a_1 z^3 + 6a_2 z^2 + 4a_3 z + a_4 \equiv f(z)$$

is a fourth- or third-degree polynomial without multiple roots, and $R(z, w)$ is a rational function of its arguments.

Diverse problems in geometry, analysis, and mechanics lead to integrals of the form (1). One of the problems of this kind was the problem of finding the length of an arc of an ellipse. It was this problem that led to the terms *elliptic integral* and *elliptic function.*

Take the ellipse

$$x = a \sin t, \qquad y = b \cos t$$

and let $c^2 = a^2 - b^2$ and $c/a = k$. For the arclength differential we have that

$$ds^2 = dx^2 + dy^2 = (a^2 \cos^2 t + b^2 \sin^2 t)\, dt^2$$
$$= (a^2 - c^2 \sin^2 t)\, dt^2 = a^2(1 - k^2 \sin^2 t)\, dt^2.$$

Therefore,

$$s = a \int \sqrt{1 - k^2 \sin^2 t} \, dt.$$

If instead of t we introduce $\xi = \sin t$, then for the arclength we get

$$s = a \int \sqrt{\frac{1 - k^2 \xi^2}{1 - \xi^2}} \, d\xi \quad \text{or} \quad s = a \int \frac{1 - k^2 \xi^2}{\sqrt{(1 - \xi^2)(1 - k^2 \xi^2)}} \, d\xi,$$

which is really a special case of (1).

We saw in §7 that with the help of a suitable linear fractional transformation it is possible to reduce the radical in the integral (1) to the form $\sqrt{4x^3 - g_2 x - g_3}$. Suppose that this linear fractional transformation has the form

$$z = \frac{\alpha x + \beta}{\gamma x + \delta} \qquad \left(\begin{vmatrix} \alpha & \beta \\ \gamma & \delta \end{vmatrix} = 1 \right).$$

Then

$$\sqrt{a_0 z^4 + 4a_1 z^3 + 6a_2 z^2 + 4a_3 z + a_4} = \frac{\sqrt{4x^3 - g_2 x - g_3}}{(\gamma x + \delta)^2},$$

and hence

$$\int R\left(z, \sqrt{a_0 z^4 + 4a_1 z^3 + 6a_2 z^2 + 4a_3 z + a_4}\right) dz$$

$$= \int R\left(\frac{\alpha x + \beta}{\gamma x + \delta}, \frac{\sqrt{4x^3 - g_2 x - g_3}}{(\gamma x + \delta)^2}\right) \frac{dx}{(\gamma x + \delta)^2}.$$

In particular,

$$\int \frac{dz}{\sqrt{a_0 z^4 + 4a_1 z^3 + 6a_2 z^2 + 4a_3 z + a_4}} = \int \frac{dz}{\sqrt{4x^3 - g_2 x - g_3}}.$$

Thus, instead of (1) we can consider the integral

$$\int R\left(z, \sqrt{4z^3 - g_2 z - g_3}\right) dz, \tag{2}$$

where R again denotes a rational function of its two arguments.

If we introduce the function $\wp(u)$ corresponding to the invariants g_2 and g_3 and let $z = \wp(u)$, then the integral (2) takes the form

$$\int R(\wp, -\wp')\wp' \, du = \int R_1(\wp, \wp') \, du, \tag{3}$$

i.e., we arrive at the integral of an elliptic function. The substitution $z = \wp(u)$ means that

$$u = \int_z^\infty \frac{dz}{\sqrt{4x^2 - g_2 x - g_3}}, \tag{4}$$

and the transition from (2) to (3) can be looked at as being for the purpose of regarding the general elliptic integral (2) as a function of the corresponding elliptic integral of the first kind (4).

To find the integral (3) it is most convenient to expand the elliptic function $R_1(\wp, \wp')$ into partial fractions:

$$R_1(\wp, \wp') = C + \sum_{k=1}^{n} A_k \zeta(u - a_k) + \sum_{\substack{k,r \\ (r \geq 2)}} A_k^{(r-1)} \wp^{(r-2)}(u - a_k).$$

Integration gives us that

$$\int R_1(\wp, \wp')\, du = C_1 + Cu + \sum_{k=1}^{n} A_k \ln \sigma(u - a)_k)$$

$$- \sum_{k=1}^{n} A_k^{(1)} \zeta(u - a_k) + \sum_{\substack{k,r \\ (r \geq 3)}} A_k^{(r-1)} \wp^{(r-3)}(u - a_k).$$

We now take into account the addition theorems for the functions ζ and $\wp$. In view of these theorems

$$\sum_{k=1}^{n} A_k^{(1)} \zeta(u - a_k) = -A\zeta(u) + R_2(\wp, \wp'),$$

$$\sum_{\substack{k,r \\ (r \geq 3)}} A_k^{(r-1)} \wp^{(r-3)}(u - a_k) = R_3(\wp, \wp'),$$

where A is a constant.

On the basis of these formulas,

$$\int R_1(\wp, \wp')\, du$$

$$= Cu + \sum_{k=1}^{n} A_k \ln \sigma(u - a_k) + A\zeta(u) + R^*(\wp, \wp').$$

Since $\sum_1^n A_k = 0$, this formula can be rewritten in the form

$$\int R_1(\wp, \wp')\, du$$

$$= Cu + A\zeta(u) + \sum_{k=1}^{n} A_k \ln \frac{\sigma(u - a_k)}{\sigma(u)} + R^*(\wp, \wp'). \tag{5}$$

The last term on the right-hand side is an elliptic function; the first three are not.

Let us pass from the variable u to the original variable $z = \wp(u)$. Then the last term on the right-had side of (5) is written in the form $R^*(z, w)$, where $w^2 = 4z^3 - g_2 z - g_3$. This is the algebraic part of the integral (2).

As for the transcendental part, it can be constructed with the help of the elements u, $\zeta(u)$, and

$$\ln \frac{\sigma(u - a)}{\sigma(u)} + u\zeta(a).$$

The first of these functions is

$$u = \int \frac{dz}{w},$$

the second is equal to

$$\zeta(u) = -\int \wp(u)\, du = -\int \frac{z\, dz}{w},$$

and the third function is

$$\ln \frac{\sigma(u - a)}{\sigma(u)} + u\zeta(a) = \int \left\{ \frac{\sigma'(u - a)}{\sigma(u - a)} - \frac{\sigma'(u)}{\sigma(u)} + \zeta(a) \right\} du$$

$$= \int \{\zeta(u - a) - \zeta(u) + \zeta(a)\}\, du$$

$$= \frac{1}{2} \int \frac{\wp'(u) + \wp'(a)}{\wp(u) - \wp(a)}\, du.$$

Introducing $z = \wp(u)$ and $w = \wp'(u)$, we let $\wp(a) = z_0$ and $\wp'(a) = w_0$. Then the third function takes the form

$$\frac{1}{2} \int \frac{w + w_0}{z - z_0} \frac{dz}{w}.$$

The integral $u = \int w^{-1}\, dz$ was called earlier an elliptic integral of the first kind, and now we call the integral

$$\int \frac{z\, dz}{w}$$

a *normal integral of the second kind*, and the integral

$$\frac{1}{2} \int \frac{w + w_0}{z - z_0} \frac{dz}{w}$$

a *normal integral of the third kind*.

Thus, *every elliptic integral is the result of adding elliptic integrals of three kinds and a certain rational function of z and w.*

This result, which we obtained with the help of the theory of elliptic functions constructed above, can be proved independently of this theory and is a special case of general theorems about reduction of elliptic and hyperelliptic integrals, i.e., integrals of the form $\int R(z, Z)\, dz$, where Z^2 is a polynomial of degree $n \geq 3$, and R denotes a rational function.

CHAPTER 4

Theta Functions

§18. Representations of theta functions by infinite products

In §3 theta functions were defined as infinite series. We now study the expansion of theta functions in infinite products.

To get these expansions we consider the function

$$f(s) = \prod_{k=1}^{\infty}(1 - h^{2k-1}s) \prod_{h=1}^{\infty}(1 - h^{2k-1}s^{-1}), \tag{1}$$

where h is a constant with modulus less than 1, and s is a complex variable. The infinite products converge absolutely for arbitrary $s \neq 0$, and the function $f(s)$ defined by (1) is obviously regular at each finite nonzero point s. Further, it follows from the form of the right-hand side of (1) that $f(s)$ satisfies the following functional equation:

$$f(s) = -hs\,f(h^2 s). \tag{2}$$

The function $f(s)$ can be expanded in a Laurent series. Suppose that this expansion has the form

$$f(s) = \sum_{k=-\infty}^{\infty} a_k s^k. \tag{3}$$

Using (2), we get that

$$\sum_{k=-\infty}^{\infty} a_k s^k = - \sum_{k=-\infty}^{\infty} a_k h^{2k+1} s^{k+1},$$

which implies that $a_k = -a_{k-1}h^{2k-1}$. This relation can be rewritten in the form

$$(-1)^k a_k h^{-k^2} = (-1)^{k-1} a_{k-1} h^{-(k-1)^2}.$$

Thus, $(-1)^k a_k h^{-k^2}$ does not depend on k, and hence

$$(-1)^k a_k h^{-k^2} = a_0.$$

49

Our expansion (3) takes the form

$$f(s) = a_0 \sum_{k=-\infty}^{\infty} (-1)^k h^{k^2} s^k;$$

from this,

$$f(e^{2\pi i v}) = a_0 \left\{ 1 + 2 \sum_{k=1}^{\infty} (-1)^k h^{k^2} \cos 2k\pi v \right\}.$$

The expression in the curly brackets is none other than $\vartheta_0(v)$. Consequently,

$$\vartheta_0(v) = f(e^{2\pi i v})/a_0.$$

On the other hand,

$$f(e^{2\pi i v}) = \prod_{k=1}^{\infty} (1 - h^{2k-1} e^{2\pi i v})(1 - h^{2k-1} e^{-2\pi i v})$$

$$= \prod_{k=1}^{\infty} (1 - 2h^{2k-1} \cos 2\pi v + h^{4k-2}).$$

Thus,

$$\vartheta_0(v) = \frac{1}{a_0} \prod_{k=1}^{\infty} (1 - 2h^{2k-1} \cos 2\pi v + h^{4k-2}).$$

We have obtained an expansion of $\vartheta_0(v)$ in an infinite product, but the numerical factor $1/a_0$ has not yet been determined.

Let us find it. To do so, we let

$$f_n(s) = \prod_{k=1}^{n} (1 - h^{2k-1} s)(1 - h^{2k-1} s^{-1}). \tag{4}$$

Multiplying this out, we get

$$f_n(s) = a_0^{(n)} + a_1^{(n)}(s + 1/s) + \cdots + a_n^{(n)}(s^n + 1/s^n).$$

Further,

$$a_n^{(n)} = (-1)^n h^{1+3+5+\cdots+(2n-1)} = (-1)^n h^{n^2}. \tag{5}$$

On the other hand, by (4),

$$(sh - h^{2n}) f_n(h^2 s) = -(1 - h^{2n+1} s) f_n(s).$$

Therefore,

$$(sh - h^{2n}) \sum_{k=-n}^{n} a_k^{(n)} h^{2k} s^k = -(1 - h^{2n+1} s) \sum_{k=-n}^{n} a_k^{(n)} s^k,$$

or

$$\sum_{k=-n}^{n} a_k^{(n)} (h^{2k+1} - h^{2n+1}) s^{k+1} = \sum_{k=-n}^{n} a_k^{(n)} (h^{2k+2n} - 1) s^k.$$

Comparison of coefficients gives us

$$a_k^{(n)}(h^{2k+1} - h^{2n+1}) = a_{k+1}^{(n)}(h^{2(k+n+1)} - 1).$$

If here we successively let $k = 0, \ldots, n - 1$ and multiply the resulting equalities, we have

$$(-1)^n a_0^{(n)} \prod_{k=1}^{n}(h^{2k-1} - h^{2n+1}) = a_n^{(n)} \prod_{k=1}^{n}(1 - h^{2(n+k)}).$$

From this, by (5),

$$a_0^{(n)} = \frac{h^{n^2} \prod_{k=1}^{n}(1 - h^{2(n+k)})}{\prod_{k=1}^{n}(h^{2k-1} - h^{2n+1})},$$

or

$$a_0^{(n)} = \frac{\prod_{k=1}^{n}(1 - h^{2(n+k)})}{\prod_{k=1}^{n}(1 - h^{2k})}.$$

The quantity a_0 we are looking for is equal to

$$a_0 = \lim_{n \to \infty} a_0^{(n)}.$$

Indeed, by virtue of the formulas determining the coefficients in the Laurent series,

$$a_0 = \frac{1}{2\pi i} \oint f(s) \frac{ds}{s}, \qquad a_0^{(n)} = \frac{1}{2\pi i} \oint f_n(s) \frac{ds}{s}$$

where the integrals are taken over the unit circle, and on it $f_n(s)$ tends to $f(s)$ uniformly as $n \to \infty$. It follows from the expression for $a_0^{(n)}$ that

$$a_0 = 1 / \prod_{k=1}^{\infty}(1 - h^{2k}).$$

Thus the final formula has the form

$$\vartheta_0(v) = H_0 \prod_{k=1}^{\infty}(1 - h^{2k-1}e^{2\pi i v})(1 - h^{2k-1}e^{-2\pi i v})$$

$$= H_0 \prod_{k=1}^{\infty}(1 - 2h^{2k-1}\cos 2\pi v + h^{4k-2}),$$

where $H_0 = \prod_{k=1}^{\infty}(1 - h^{2k})$. From this it is now not hard to get analogous expansions for the remaining theta functions. They are all contained in Table IX. For example, we derive the expansion of $\vartheta_1(v)$. With this goal we use the equality

$$\vartheta_1(v) = \frac{1}{i} h^{1/4} e^{\pi i v} \vartheta_0(v + \tau/2).$$

This implies that

$$\vartheta_1(v) = \frac{1}{i} H_0 h^{1/4} e^{\pi i v} \prod_{k=1}^{\infty} (1 - h^{2k} e^{2\pi i v})(1 - h^{2k-2} e^{-2\pi i v})$$

$$= \frac{1}{i} H_0 h^{1/4} e^{\pi i v} (1 - e^{-2\pi i v})$$

$$\times \prod_{k=1}^{\infty} (1 - h^{2k} e^{2\pi i v})(1 - h^{2k} e^{-2\pi i v})$$

$$= 2 H_0 h^{1/4} \sin \pi v \prod_{k=1}^{\infty} (1 - 2h^{2k} \cos 2\pi v + h^{4k}).$$

With the infinite-product expansions of the theta functions in hand it is not hard to write the collection of all zeros of these functions, as well as to obtain the values of these functions at zero([15]) and, in particular, to prove that $\vartheta_1'(0) = \pi \vartheta_0(0)\vartheta_2(0)\vartheta_3(0)$ (all this is contained in Table IX).

§19. The connection between sigma functions and theta functions

We compare the function $\sigma(u)$ with the function $\vartheta_1(u/2\omega)$. The zeros of each of these functions are simple and have the form

$$u = 2m\omega + 2m'\omega' \qquad (m, m' = 0, \pm 1, \pm 2, \dots).$$

We consider the expression

$$f(u) = \frac{e^{\alpha u^2} \sigma(u)}{\vartheta_1(u/2\omega)}.$$

This is a function not having singular points at a finite distance, because the zeros of the denominator are zeros of the same multiplicity for the numerator.

We find $f(u + 2\omega)$ and $f(u + 2\omega')$:

$$f(u + 2\omega) = e^{\alpha(u+2\omega)^2} \frac{\sigma(u + 2\omega)}{\vartheta_1(u/2\omega + 1)}$$

$$= e^{\alpha u^2} e^{4\omega\alpha(u+\omega)} e^{2\eta(u+\omega)} \frac{\sigma(u)}{\vartheta_1(u/2\omega)}$$

$$= e^{2(2\omega\alpha+\eta)(u+\omega)} f(u),$$

$$f(u + 2\omega') = e^{\alpha(u+2\omega')^2} \frac{\sigma(u + 2\omega')}{\vartheta_1(u/2\omega + \tau)}$$

$$= e^{\alpha u^2} e^{4\alpha\omega(u+\omega')} \frac{e^{2\eta'(u+\omega')}}{h^{-1} e^{-2\pi i u/2\omega}} \frac{\sigma(u)}{\vartheta_1(u/2\omega)}$$

$$= e^{2(2\omega'\alpha+\eta')+\pi i/\omega\}(u+\omega')} f(u)'.$$

([15]) We call them the *zero values*.

Taking into account the equality $\eta\omega' - \eta'\omega = \pi i/2$, we get

$$2(2\omega'\alpha + \eta') + \pi i/\omega = 2\tau(2\omega\alpha + \eta).$$

If we thus let $\alpha = -\eta/2\omega$, the above equalities take the form

$$f(u + 2\omega) = f(u), \qquad f(u + 2\omega') = f(u).$$

But since $f(u)$ is an entire function, it becomes a constant for the indicated choice of α. Consequently,

$$\sigma(u) = Ce^{\eta u^2/2\omega}\vartheta_1(u/2\omega).$$

To determine the constant C we differentiate the equality written and let $u = 0$. This gives us that

$$1 = C \cdot \frac{1}{2\omega}\vartheta_1'(0),$$

from which $C = 2\omega/\vartheta_1'(0)$, and hence

$$\sigma(u) = \frac{2\omega e^{\eta u^2/2\omega}\vartheta_1(u/2\omega)}{\vartheta_1'(0)}. \tag{1}$$

Analogous relations (they are presented in Table X) hold for the remaining sigma and theta functions.

Relation (1) enables us to use a theta function instead of a sigma function for a representation of an arbitrary elliptic function from its zeros and poles.

Let $f(u)$ be an elliptic function, and let $a_1,\ldots,a_n$ be its poles and $b_1,\ldots,b_n$ its zeros in a fundamental parallelogram. Suppose, further, that

$$a_1 + \cdots + a_n = b_1^* + b_2 + \cdots + b_n. \tag{2}$$

We have seen that

$$f(u) = C_1\frac{\sigma(u - b_1^*)\sigma(u - b_2)\cdots\sigma(u - b_n)}{\sigma(u - a_1)\sigma(u - a_2)\cdots\sigma(u - a_n)},$$

where C_1 is a constant.

We now obtain the representation

$$f(u) = C_1 e^{\frac{n}{2\omega}\{(u-b_1^*)^2+\cdots+(u-b_n)^2-(u-a_1)^2-\cdots-(u-a_n)^2\}}$$

$$\times \frac{\vartheta_1\left(\dfrac{u - b_1^*}{2\omega}\right)\cdots\vartheta_1\left(\dfrac{u - b_n}{2\omega}\right)}{\vartheta_1\left(\dfrac{u - a_1}{2\omega}\right)\cdots\vartheta_1\left(\dfrac{u - a_n}{2\omega}\right)}.$$

In the expression

$$(u - b_1^*)^2 + \cdots + (u - b_n)^2 - (u - a_1)^2 - \cdots - (u - a_n)^2$$

the terms with u^2 cancel out. By (2), the same happens for the terms containing u to the first power. Thus,

$$f(u) = C \frac{\vartheta_1\left(\dfrac{u - b_1^*}{2\omega}\right) \vartheta_1\left(\dfrac{u - b_2}{2\omega}\right) \cdots \vartheta_1\left(\dfrac{u - b_n}{2\omega}\right)}{\vartheta_1\left(\dfrac{u - a_1}{2\omega}\right) \vartheta_1\left(\dfrac{u - a_2}{2\omega}\right) \cdots \vartheta_1\left(\dfrac{u - a_n}{2\omega}\right)}.$$

§20. Expansion of the functions $\zeta(u)$ and $\wp(u)$ in simple series

We turn to the formula

$$\sigma(u) = 2\omega e^{2\eta\omega v^2} \frac{\vartheta_1(v)}{\vartheta_1'(0)},$$

where $v = u/(2\omega)$. Taking the logarithmic derivative of both sides with respect to u, we get

$$\zeta(u) = \frac{\eta}{\omega} u + \frac{1}{2\omega} \frac{\vartheta_1'(v)}{\vartheta_1(v)}.$$

Let us now replace $\vartheta_1(v)$ by its expansion in an infinite product. This gives us the following expansion of $\zeta(u)$:

$$\zeta(u) = \frac{\eta}{\omega} u + \frac{\pi}{2\omega} \left\{ \cot \pi v + \sum_{k=1}^{\infty} \frac{4h^{2k} \sin 2\pi v}{1 - 2h^{2k} \cos 2\pi v + h^{4k}} \right\}. \tag{1}$$

Here we have a simple infinite series, in contrast to the double series in the definition of $\zeta(u)$. The series obtained can be represented in the following form, which is more convenient for many purposes:

$$\zeta(u) = \frac{\eta}{\omega} u + \frac{\pi i}{2\omega} \left\{ \frac{z + z^{-1}}{z - z^{-1}} + \sum_{k=1}^{\infty} \left(\frac{2h^{2k} z^{-2}}{1 - h^{2k} z^{-2}} - \frac{2h^{2k} z^2}{1 - h^2 z^2} \right) \right\}, \tag{2}$$

where $z = e^{\pi i u/2\omega}$.

To get an analogous expansion of $\wp(u)$ we differentiate (2) with respect to u. This gives us

$$\wp(u) = -\frac{\eta}{\omega} - \left(\frac{\pi}{\omega}\right)^2 \left\{ \frac{1}{(z - z^{-1})^2} \right.$$
$$\left. + \sum_{k=1}^{\infty} \left[\frac{h^{2k} z^{-2}}{(1 - h^{2k} z^{-2})^2} + \frac{h^{2k} z^2}{(1 - h^{2k} z^2)^2} \right] \right\}. \tag{3}$$

We use this series to express η, e_1, e_2, and e_3 in terms of h. Setting $u = \omega$ in (3), and hence $z = i$, we have

$$e_1 = \frac{\eta}{\omega} + \left(\frac{\pi}{\omega}\right)^2 \left\{ \frac{1}{4} + 2\sum_{k=1}^{\infty} \frac{h^{2k}}{(1 - h^{2k})^2} \right\}.$$

Similarly, setting $u = \omega'$, and thus $z = h^{1/2}$, we have

$$e_3 = -\frac{\eta}{\omega} - 2\left(\frac{\pi}{\omega}\right)^2 \sum_{k=1}^{\infty} \frac{h^{2k-1}}{(1 - h^{2k-1})^2}.$$

Finally, setting $u = -\omega - \omega'$, and thus $z = -ih^{-1/2}$, we have

$$e_2 = -\frac{\eta}{\omega} + 2\left(\frac{\pi}{\omega}\right)^2 \sum_{k=1}^{\infty} \frac{h^{2k-1}}{(1 + h^{2k-1})^2}.$$

Adding these equalities termwise and using the equality $e_1 + e_2 + e_3 = 0$, we find after simple transformations that

$$\eta\omega = \frac{\pi^2}{12}\left\{1 - 24\sum_{k=1}^{\infty} \frac{h^{2k}}{(1 - h^{2k})^2}\right\}.$$

§21. Expressions for e_1, e_2, and e_3 in terms of the zero values of the theta functions

We recall the formula

$$\sqrt{\wp(u) - e_\alpha} = \sigma_\alpha(u)/\sigma(u) \qquad (\alpha = 1, 2, 3).$$

Setting $u = \omega_\beta$ here, we get

$$\sqrt{e_\beta - e_\alpha} = \sigma_\alpha(\omega_\beta)/\sigma(\omega_\beta).$$

Let us express the right-hand side in terms of theta functions. For example, taking $\beta = 1$ and $\alpha = 2$, we have

$$\sqrt{e_1 - e_2} = \frac{1}{2\omega}\frac{\vartheta_0(0)\vartheta_1'(0)}{\vartheta_2(0)\vartheta_3(0)} = \frac{1}{2\omega}\frac{\vartheta_0\vartheta_1'}{\vartheta_2\vartheta_3}.$$

Here, as everywhere below, ϑ_0, ϑ_2, ϑ_3 and ϑ_1' represent the values at $v = 0$ of the functions $\vartheta_0(v)$, $\vartheta_2(v)$, $\vartheta_3(v)$, and $\vartheta_1'(v)$. But since (see §18 and Table IX)

$$\vartheta_1' = \pi\vartheta_0\vartheta_2\vartheta_3, \tag{1}$$

it follows that

$$\sqrt{e_1 - e_2} = \pi\vartheta_0^2/2\omega.$$

Similarly,

$$\sqrt{e_2 - e_1} = i\pi\vartheta_0^2/2\omega,$$

and

$$\sqrt{e_2 - e_3} = -i\sqrt{e_3 - e_2} = -\pi\vartheta_2^2/2\omega,$$
$$\sqrt{e_3 - e_1} = -i\sqrt{e_1 - e_3} = -i\pi\vartheta_3^2/2\omega.$$

It follows from these formulas that

$$\left.\begin{aligned}
e_1 - e_2 &= (\pi/2\omega)^2\vartheta_0^4, \\
e_2 - e_3 &= (\pi/2\omega)^2\vartheta_2^4, \\
e_1 - e_3 &= (\pi/2\omega)^2\vartheta_3^4.
\end{aligned}\right\} \tag{2}$$

From this,

$$\vartheta_3^4 = \vartheta_0^4 + \vartheta_2^4. \tag{3}$$

An important role in what follows is played by the function

$$\lambda = \lambda(\tau) \equiv \vartheta_2^4(0|\tau)/\vartheta_3^4(0|\tau). \tag{4}$$

If we use (2) and (3) and recall formula (6) in §6, then we get the following representation of the modular function $J(\tau)$ in terms of the functions $\lambda(\tau)$:

$$J = \frac{4(\lambda^2 - \lambda + 1)^3}{27\lambda^2(1 - \lambda)^2}. \tag{5}$$

§22. Transformation of theta functions

In considering theta functions we have so far studied their dependence on the argument v and have not directed our attention to the dependence on the parameter τ nor on $h = e^{\pi i \tau}$ ($\Im\tau > 0$). In particular, we investigated how a theta function changes when one of the periods or half-periods is added to the argument v.

We now study the dependence of theta functions on the parameter τ. Appearing here instead of the group of translations (by periods or half-periods) is the modular group Σ of all substitutions

$$\tau^* = \frac{\alpha\tau + \beta}{\gamma\tau + \delta},$$

where α, β, γ, and δ are integers such that $\alpha\delta - \beta\gamma = 1$. The group Σ, as established above (see §9), is generated by the two basic substitutions

$$S = \begin{pmatrix} 1 & 1 \\ 0 & 1 \end{pmatrix}, \qquad T = \begin{pmatrix} 0 & 1 \\ -1 & 0 \end{pmatrix}.$$

Therefore, it suffices to investigate how theta functions transform when τ is subjected to these two basic transformations.

The transition from τ to $\tau + 1$ corresponds to replacement of h by $-h$, and the corresponding transformation formulas are obtained very simply by expanding theta functions in series. These formulas have the form

$$\left.\begin{aligned}
\vartheta_1(v|\tau + 1) &= i^{1/2}\vartheta_1(v|\tau), \\
\vartheta_2(v|\tau + 1) &= i^{1/2}\vartheta_2(v|\tau), \\
\vartheta_3(v|\tau + 1) &= \vartheta_0(v|\tau), \\
\vartheta_0(v|\tau + 1) &= \vartheta_3(v|\tau).
\end{aligned}\quad (i^{1/2} = e^{\pi i/4})\right\} \tag{1}$$

We now investigate the transition from τ to $-1/\tau$ and introduce for convenience the notation $\tau' = -1/\tau$. Let

$$f(v) = e^{\pi i \tau' v^2} \vartheta_3(\tau' v | \tau') / \vartheta_3(v | \tau).$$

It is not hard to verify that this function does not have singularities. Indeed, the only zeros (and they are simple) of the denominator are the points

$$v = (m + 1/2)\tau + (n + 1/2), \tag{2}$$

where m and n are integers. At the same time, these points are zeros of the numerator, because the latter vanishes for

$$\tau' v = (m' + 1/2)\tau' + (n' + 1/2),$$

where m' and n' are integers, i.e., for

$$v = (m' + 1/2) - (n' + 1/2)\tau.$$

Thus, $f(v)$ is an entire transcendental function. But it is easy to verify that $f(v)$ has periods 1 and τ. Consequently, $f(v)$ is a constant, i.e.,

$$\vartheta_3(\tau' v | \tau') = A e^{-\pi i \tau' v^2} \vartheta_3(v | \tau). \tag{3}$$

If we replace v here by $v + 1/2$, $v - \tau/2$ and $v + (1 - \tau)/2$, and use the relations in §3, we get the following formulas:

$$\left. \begin{array}{l} \vartheta_2(\tau' v | \tau') = A e^{-\pi i \tau' v^2} \vartheta_0(v | \tau), \\ \vartheta_0(\tau' v | \tau') = A e^{-\pi i \tau' v^2} \vartheta_2(v | \tau), \\ \vartheta_1(\tau' v | \tau') = i A e^{-\pi i \tau' v^2} \vartheta_1(v | \tau) \end{array} \right\} \tag{3'}$$

and everything reduces to finding the constant A. With this goal we write these formulas for $v = 0$, and we first differentiate the last formula with respect to v. This gives us that

$$\vartheta_3(0 | \tau') = A \vartheta_3(0 | \tau),$$
$$\vartheta_2(0 | \tau') = A \vartheta_0(0 | \tau),$$
$$\vartheta_0(0 | \tau') = A \vartheta_2(0 | \tau),$$
$$\tau' \vartheta_1'(0 | \tau') = i A \vartheta_1'(0 | \tau).$$

We now consider that $\vartheta_1' = \pi \vartheta_0 \vartheta_2 \vartheta_3$. In view of this relation,

$$\tau' \pi \vartheta_0(0 | \tau') \vartheta_2(0 | \tau') \vartheta_3(0 | \tau') = i A \pi \vartheta_0(0 | \tau) \vartheta_2(0 | \tau) \vartheta_3(0 | \tau),$$

and hence $A^2 \tau' = i$, from which $A^2 = -i\tau$, so that

$$A = \pm \sqrt{-i\tau}, \tag{3''}$$

where the radical is understood to be the value of it with positive real part.

It now remains to determine the sign in $(3'')$, i.e., in the equality

$$\vartheta_3(0|\tau') = \pm\sqrt{-i\tau}\,\vartheta_3(0|\tau). \tag{4}$$

Both the quantities

$$\vartheta_3(0|\tau'), \qquad \sqrt{-i\tau}\,\vartheta_3(0|\tau) \tag{5}$$

are regular functions of τ in the upper half-plane. If τ has a purely imaginary value, then $h = e^{\pi i \tau}$ and $h' = e^{\pi i \tau'}$ are positive, and the quantities (5) are also positive, as follows from the definition of $\vartheta_3(v)$ by means of a trigonometric series. We see that we have to take the plus sign in (4). Thus, (3) and $(3')$ take the form

$$\left.\begin{array}{l}
\vartheta_3(\tau'v|\tau') = \sqrt{-i\tau}\,e^{-\pi i \tau' v^2}\vartheta_3(v|\tau), \\[4pt]
\vartheta_2(\tau'v|\tau') = \sqrt{-i\tau}\,e^{-\pi i \tau' v^2}\vartheta_0(v|\tau), \\[4pt]
\vartheta_0(\tau'v|\tau') = \sqrt{-i\tau}\,e^{-\pi i \tau' v^2}\vartheta_2(v|\tau), \\[4pt]
\vartheta_1(\tau'v|\tau') = i\sqrt{-i\tau}\,e^{-\pi i \tau' v^2}\vartheta_1(v|\tau).
\end{array}\right\} \tag{6}$$

§23. The modular function $\lambda(\tau)$

This function was introduced at the end of §21. We recall its definition:

$$\lambda = \frac{\vartheta_2^4(0|\tau)}{\vartheta_3^4(0|\tau)} \equiv \lambda(\tau).$$

On the basis of (1) in §22,

$$\lambda(\tau + 1) = -\frac{\vartheta_2^4(0|\tau)}{\vartheta_0^4(0|\tau)}.$$

But since, by (3) in §21,

$$\vartheta_0^4(0|\tau) = \vartheta_3^4(0|\tau) - \vartheta_2^4(0|\tau),$$

it follows that

$$\lambda(\tau + 1) = \frac{1}{1 - 1/\lambda(\tau)},$$

or

$$\lambda(\tau + 1) = \frac{\lambda(\tau)}{\lambda(\tau) - 1}. \tag{1}$$

Similarly, in view of (6) in §22,

$$\lambda\left(-\frac{1}{\tau}\right) = \frac{\vartheta_0^4(0|\tau)}{\vartheta_3^4(0|\tau)}. \tag{2'}$$

Consequently,

$$\lambda(-1/\tau) = 1 - \lambda(\tau). \tag{2}$$

Formulas (1) and (2) show that $\lambda(\tau)$ is not invariant with respect to all transformations in the modular group. However, we can determine a certain subgroup of the full modular group Σ (to be denoted by Σ_2) with respect to which $\lambda(\tau)$ is invariant. Therefore, $\lambda(\tau)$ is also called a modular function.

An arbitrary substitution in the modular group can be obtained by composition of the basic substitutions $S: S\tau = \tau + 1$ and $T: T\tau = -1/\tau$. In connection with (1) and (2) this circumstance enables us to establish how $\lambda(\tau)$ changes under transformations of the full modular group Σ.

Indeed, in view of (1) and (2) we immediately have

$$\lambda(I\tau) = \lambda(\tau), \qquad \lambda(S\tau) = \frac{\lambda(\tau)}{\lambda(\tau) - 1}, \qquad \lambda(T\tau) = 1 - \lambda(\tau);$$

further,

$$\lambda(ST\tau) = \frac{\lambda(T\tau)}{\lambda(T\tau) - 1} = \frac{\lambda(\tau) - 1}{\lambda(\tau)}$$

and, analogously,

$$\lambda(TS\tau) = 1/(1 - \lambda(\tau)),$$
$$\lambda(STS\tau) = \lambda(TST\tau) = 1/\lambda(\tau).$$

We prove that this exhausts the collection of all values taken by $\lambda(U\tau)$ when U runs through Σ. With this goal we remark in the first place that

$$\lambda(S^2\tau) = \frac{\lambda(S\tau)}{\lambda(S\tau) - 1}$$
$$= \frac{\lambda(\tau)}{[\lambda(\tau) - 1][\lambda(\tau)/(\lambda(\tau) - 1) - 1]} = \lambda(\tau).$$

But since every substitution U in Σ has the form

$$U = TS^i TS^k \cdots TS^m TS^n,$$

it follows that to get the different values of the function $\lambda(U\tau)$ we need consider only the substitutions U for which each of the numbers $i, k, \ldots, m, n$ is 0 or 1. Furthermore, since $T^2 = I$, we can confine ourselves to substitutions of the form

$$TSTS\cdots T, \qquad TSTS\cdots TS, \qquad STS\cdots T, \qquad STS\cdots TS. \qquad (3)$$

But it is not hard to verify that

$$STSTST = TSTSTS = I.$$

Consequently, the substitutions $STSTST$ and $TSTSTS$ leave $\lambda(\tau)$ invariant. Therefore, there remain only the substitutions that are products of the form (3) with at most five factors, i.e., there remain

$$\left.\begin{array}{ll} S, & T \\ ST, & TS \\ STS, & TST \end{array}\right\} \quad (\alpha)$$

$$\left.\begin{array}{ll} TSTS, & STST \\ TSTST, & STSTS \end{array}\right\} \quad (\beta)$$

The substitutions (α) were considered above, As for (β), they do not give anything new; for example,

$$\lambda(TSTS\tau) = \lambda(TSTSTSST\tau) = \lambda(ST\tau).$$

Thus, the assertion is proved. We remark that the sextuple of numbers

$$\lambda, \quad \frac{1}{\lambda}, \quad 1-\lambda, \quad \frac{1}{1-\lambda}, \quad \frac{\lambda}{\lambda-1}, \quad \frac{\lambda-1}{\lambda}$$

is obtained from λ by linear substitutions in the so-called *anharmonic group*.

It is easy to construct a fundamental region of the group Σ_2. This region, call it D_2, consists of six regions equivalent to the fundamental region D of the full modular group Σ. To construct D_2 we take instead of D the equivalent region I (Figure 7) bounded by the lines

$$\Re\tau = -1, \quad \Re\tau = 0$$

and the circles

$$|\tau + 1| = 1, \qquad |\tau| = 1.$$

We then subject each point of this region to the transformation S. We say briefly that the region I is subjected to the transformation S. The region obtained is called S. The regions T, TS, $S^{-1}T$, and STS are constructed in a similar way (see Figure 7). As a result we obtain the fundamental region D_2 of the group Σ_2. It is bounded by the lines

$$\Re\tau = -1, \qquad \Re\tau = 1$$

and the circles

$$|\tau + 1/2| = 1/2, \qquad |\tau - 1/2| = 1/2,$$

and of each pair of boundaries the left-hand one is included in D_2.

The substitutions connecting the boundaries of D_2 have the form

$$\tau^* = \tau + 2 \qquad \begin{pmatrix} 1 & 2 \\ 0 & 1 \end{pmatrix},$$

$$\tau^* = \frac{\tau}{2\tau + 1} \qquad \begin{pmatrix} 1 & 0 \\ 2 & 1 \end{pmatrix}.$$

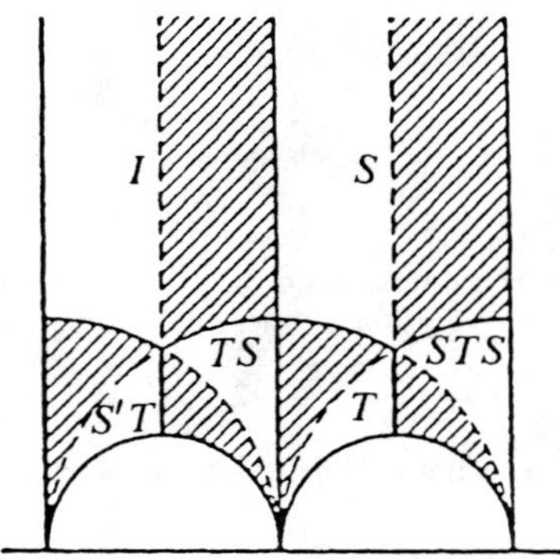

FIGURE 7

These two substitutions generate the group Σ_2, just as S and T generate Σ.

We remark that Σ_2 is completely characterized by the following property of its substitutions $\left(\begin{smallmatrix} \alpha & \beta \\ \gamma & \delta \end{smallmatrix}\right)$. First, α, β, γ, and δ are integers with $\alpha\delta - \beta\gamma = 1$, and second,

$$\alpha \equiv 1 \quad (\mathrm{mod}\ 2), \quad \beta \equiv 0 \quad (\mathrm{mod}\ 2),$$
$$\gamma \equiv 0 \quad (\mathrm{mod}\ 2), \quad \delta \equiv 1 \quad (\mathrm{mod}\ 2);$$

in other words, α and δ are odd, while β and γ are even.

The proof of these facts is left to the reader.

The domain D_2 is also called a fundamental region of the function $\lambda(\tau)$.

The following theorem holds. *For an arbitrary finite a different from 0 and 1, the equation*

$$\lambda(\tau) - a = 0 \tag{4}$$

has one and only one solution in D_2.

The proof uses formula (5) in §21:

$$J(\tau) = \frac{4(\lambda^2 - \lambda + 1)^3}{27\lambda^2(1 - \lambda)^2}. \tag{5}$$

Setting $\lambda = a$ in this formula, we get the equality

$$J(\tau) = c. \tag{6}$$

The first part of (5) is invariant under substitutions in the anharmonic group. Therefore, the same value c is obtained for the whole sextuple

$$a, \quad \frac{1}{a}, \quad 1 - a, \quad \frac{1}{1-a}, \quad \frac{a}{a-1}, \quad \frac{a-1}{a}. \tag{7}$$

If for a given a all the numbers (7) are distinct, then (6), by what was proved in §10, has precisely six simple roots interior to D_2. At these points of D_2, which are equivalent with respect to the full modular group, the function $\lambda(\tau)$ takes the sextuple of values (7) in view of what has been proved in this section. Consequently, $\lambda(\tau)$ takes the value a at one of these points.

It remains to consider what happens when the numbers in (7) are not all distinct. This is the case only when a has one of the following values:

$$-1, \quad \frac{1}{2}, \quad 2, \quad \frac{1}{2} + i\frac{\sqrt{3}}{2}, \quad \frac{1}{2} - i\frac{\sqrt{3}}{2}.$$

The corresponding values of c are 1, 1, 1, 0, 0. We now recall that in view of considerations in §10 the equation $J(\tau) = 1$ has a double root at each point τ equivalent to the point $\tau = i$ with respect to the full modular group. Of all these points, D_2 contains

$$\tau = -1 + i, \ -1/2 + i/2, \ i. \tag{8}$$

The function $\lambda(\tau)$ takes different values at them. On the other hand,

$$\frac{4(\lambda^2 - \lambda + 1)^3}{27\lambda^2(1 - \lambda)^2} = 1 + \frac{(\lambda + 1)^2(2\lambda - 1)^2(\lambda - 2)^2}{27\lambda^2(1 - \lambda)^2}.$$

From this it is now easy to see that each of the equations

$$\lambda(\tau) = -1, \quad \lambda(\tau) = 1/2, \quad \lambda(\tau) = 2$$

has in D_2 precisely one (simple) root, and these roots coincide with the numbers (8).

It is established similarly that the equations

$$\lambda(\tau) = \frac{1}{2} + i\frac{\sqrt{3}}{2}, \quad \lambda(\tau) = \frac{1}{2} - i\frac{\sqrt{3}}{2}$$

have in D_2 one (simple) root each:

$$\tau = \frac{1}{2} + i\frac{\sqrt{3}}{2}, \quad \tau = -\frac{1}{2} + i\frac{\sqrt{3}}{2}.$$

In conclusion we dwell on the correspondence of the boundaries under the mapping produced by the function $\lambda = \lambda(\tau)$.

First of all we take the representation

$$\lambda = 1 - \prod_{k=1}^{\infty} \left(\frac{1 - h^{2k-1}}{1 + h^{2k-1}}\right)^8,$$

which is obtained with the help of (2'), (2), and the infinite products for the theta functions. Here $h = e^{\pi i \tau}$. Therefore, if τ runs through the positive half CO of the imaginary axis, then h increases monotonically from 0 to 1, and hence λ also increases monotonically from 0 to 1 (Figure 8). Now take the line CB; on this line $\tau = 1 + i\eta$, where η varies from ∞ to 0. Consequently, on CB

$$\lambda = 1 - \prod_{k=1}^{\infty} \left[\frac{1 + e^{-2\pi\eta(2k-1)}}{1 - e^{-2\pi\eta(2k-1)}}\right]^8,$$

and hence as τ runs through CB the quantity λ varies from 0 to $-\infty$. The same change is experienced by λ as τ runs through the line CA.

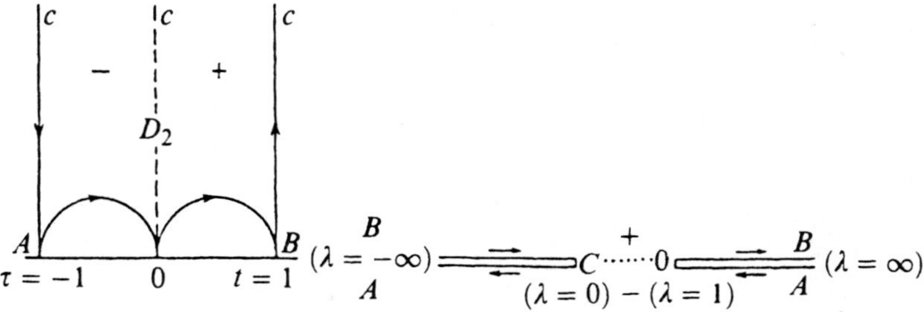

FIGURE 8

Finally, we consider the semicircle OA. If we let $\tau = -1/\tau'$, then the point τ describes the arc OA in the positive direction as the point τ' describes the line CB from C to B. We now consider (see §22) that

$$\vartheta_3(0|\tau) = (-i\tau)^{-1/2}\vartheta_3(0|-1/\tau),$$
$$\vartheta_0(0|\tau) = (-i\tau)^{-1/2}\vartheta_2(0|-1/\tau).$$

Therefore,

$$\lambda = 1 - \frac{\vartheta_2^4(0|\tau')}{\vartheta_3^4(0|\tau')},$$

from which

$$\lambda = \frac{\vartheta_0^4(0|\tau')}{\vartheta_3^4(0|\tau')},$$

and hence

$$\lambda = \prod_{k=1}^{\infty}\left(\frac{1 - h'^{2k-1}}{1 + h'^{2k-1}}\right)^8,$$

where $h' = e^{\pi i \tau'}$. This formula shows that λ runs through the real semi-axis from $\lambda = 1$ to $\lambda = \infty$ as τ runs through the arc OA in the positive direction. The same semi-axis is obtained as τ runs through the arc OB from O to B.

It follows from the above that the function $\lambda = \lambda(\tau)$ maps the right half of D_2 onto the upper half of the λ-plane, and the left half of D_2 onto the lower half of the λ-plane. The whole region D_2 is mapped onto the λ-plane, cut along the intervals $(-\infty, 0)$ and $(1, \infty)$. Adjoining to D_2 parts of the boundary in the way stipulated above, we must adjoin the lower edges of the cuts to the region in the λ-plane.

CHAPTER 5

Jacobi Functions

§24. The elliptic integral of the first kind in the forms of Jacobi and Riemann

Instead of the elliptic integral

$$u = \int_y^\infty \frac{ds}{\sqrt{4s^3 - g_2 s - g_3}}, \tag{1}$$

whose inverse is the Weierstrass function $y = \wp(u)$, the integral appearing in the Jacobi theory is

$$w = \int_0^x \frac{dt}{\sqrt{(1 - t^2)(1 - k^2 t^2)}}, \tag{2}$$

which contains only the one parameter k; this parameter is called the *modulus* of the integral.

If we let $x^2 = \xi$ and $t^2 = z$, then the integral (2) takes the form

$$w = \int_0^\xi \frac{dz}{2\sqrt{z(1 - z)(1 - k^2 z)}}. \tag{3}$$

This is the Riemann form.

The fact that the integral in Jacobi form or Riemann form contains only one parameter, and not two like the Weierstrass integral, is very convenient for various computations. The Weierstrass form is almost always preferable for theoretical considerations.

An independent treatment of the integral (3) is unnecessary, since the Weierstrass theory has been constructed. It is simpler to use the fact that there must be a linear fractional transformation of the variable z into the variable s (and hence of ξ into y) after which the integral w becomes the integral u to within a constant factor. This transformation must have the form $z = \mu/(s - \lambda)$, since z must be equal to zero for $s = \infty$. The values $z = 1, 1/k^2, \infty$ must correspond to the roots $s = e_1, e_2, e_3$ of the polynomial $4s^3 - g_2 s - g_3$. Therefore, λ must be equal to one of the numbers e_α. Let

$\lambda = e_3$ and $\mu = e_1 - e_3$. Then the root $s = e_1$ passes into $z = 1$. Accordingly,

$$z = \frac{e_1 - e_3}{s - e_3} \tag{4}$$

and hence

$$\xi = \frac{e_1 - e_3}{y - e_3}.$$

By (4),

$$s - e_3 = \frac{e_1 - e_3}{z}, \qquad s - e_1 = \frac{(e_1 - e_3)(1 - z)}{z},$$

$$s - e_2 = \frac{(e_1 - e_3)(1 - (e_2 - e_3)z/(e_1 - e_3))}{z}, \qquad ds = -\frac{(e_1 - e_3)\, dz}{z^2}.$$

Consequently,

$$u = \frac{1}{\sqrt{e_1 - e_3}} \int_0^\xi \frac{dz}{2\sqrt{z(1 - z)(1 - (e_2 - e_3)z/e_1 - e_3)}}.$$

To identify this formula with (3) it remains to let

$$k^2 = \frac{e_2 - e_3}{e_1 - e_3} \tag{5}$$

and

$$w = \sqrt{e_1 - e_3}\, u.$$

Our result can be formulated as follows: given a quantity k^2 (finite and different from 0 and 1), take some e_1, e_2, e_3, with sum zero such that (5) holds; construct the corresponding function $\wp(u)$, and then

$$\xi = \frac{e_1 - e_3}{\wp(w/\sqrt{e_1 - e_3}) - e_3}.$$

The inverse of the integral (2) has the form

$$x = \frac{\sqrt{e_1 - e_3}}{\sqrt{\wp(w/\sqrt{e_1 - e_3}) - e_3}}. \tag{6}$$

We express this function in terms of theta functions. The formulas needed for this are contained in Tables VI and X. Our result will have the form

$$x = \frac{\pi}{2\omega} \vartheta_3^2 \frac{\sigma(w/\sqrt{e_1 - e_3})}{\sigma_3(w/\sqrt{e_1 - e_3})}$$

$$= \frac{\pi \vartheta_0 \vartheta_3^2}{\vartheta_1'} \frac{\vartheta_1(w/2\omega\sqrt{e_1 - e_3})}{\vartheta_0(w/2\omega\sqrt{e_1 - e_3})}$$

$$= \frac{\vartheta_3}{\vartheta_2} \frac{\vartheta_1(w/2\omega\sqrt{e_1 - e_3})}{\vartheta_0(w/2\omega\sqrt{e_1 - e_3})}.$$

The integral (2) was studied as a function of x and k^2 as far back as Legendre. Special attention was given to the so-called normal case, when k^2 is positive and less than 1 and x lies in $[0, 1]$. In this case it is natural to set $t = \sin \psi$ and $x = \sin \varphi$. Then the integral (2) takes the form

$$w = \int_0^\varphi \frac{d\psi}{\sqrt{1 - k^2 \sin^2 \psi}}. \tag{2^{bis}}$$

Inverting this integral, Jacobi called φ the amplitude of w, writing $\varphi = \operatorname{am} w$. Then the result of inverting (2) is

$$x = \sin \varphi = \sin \operatorname{am} w.$$

Jacobi called this function the sine of the amplitude (*sinus amplitudinis*). Further,

$$\sqrt{1 - x^2} = \cos \varphi = \cos \operatorname{am} w. \tag{7}$$

Moreover, Jacobi introduced also the function

$$\sqrt{1 - k^2 x^2} = \Delta \varphi = \Delta \operatorname{am} w, \tag{8}$$

which was called the delta of the amplitude. Both functions (7) and (8) become equal to 1 for $w = 0$. The Jacobi notation is not used at the present time; it has been replaced by the Gudermann notation

$$x = \operatorname{sn} w, \qquad \sqrt{1 - x^2} = \operatorname{cn} w, \qquad \sqrt{1 - k_-^2 x^2} = \operatorname{dn} w.$$

§25. The Jacobi functions

In §24 we introduced the function

$$\operatorname{sn} w = \frac{\vartheta_3}{\vartheta_2} \frac{\vartheta_1(w/2\omega\sqrt{e_1 - e_3})}{\vartheta_0(w/2\omega\sqrt{e_1 - e_3})}.$$

This is a meromorphic function of w that further depends on the single parameter $h = e^{\pi i \tau}$ alone, since h is the only quantity on which $2\omega\sqrt{e_1 - e_3} = \pi\vartheta_3^2$ as well as the coefficients of the theta functions depend. However, it follows from considerations in §24 that the modulus k can be taken instead of $h = e^{\pi i \tau}$ or τ as the parameter on which the Jacobi functions depend. Indeed, it follows from §24 that, given k, we can determine a number τ in the upper half-plane in such a way that the function $x = \operatorname{sn} w$ constructed for τ by means of the function $\wp$ or the theta functions is the inverse of the integral

$$w = \int_0^x \frac{dt}{\sqrt{(1 - t^2)(1 - k^2 t^2)}}.$$

Thus, together with $\operatorname{sn}(w|\tau)$ a more complete notation for $\operatorname{sn} w$ is $\operatorname{sn}(w;k)$. A similar remark applies to the functions $\operatorname{cn} w$ and $\operatorname{dn} w$.

The numbers 2 and τ are primitive periods for the ratio $\vartheta_1(v)/\vartheta_0(v)$.

Hence, $4\omega\sqrt{e_1 - e_3}$ and $2\omega'\sqrt{e_1 - e_3}$ are primitive periods for the function $\operatorname{sn} w$. These are functions of h. The following notation is used:

$$\omega\sqrt{e_1 - e_3} = K, \qquad \omega'\sqrt{e_1 - e_3} = iK'.$$

Thus, $4K$ and $2iK'$ are primitive periods of $\operatorname{sn} w$.

It is inconvenient always to write $w/2K$ as the argument of the theta functions.

Following Riemann, we introduce the notation

$$\theta_\alpha(w) = \vartheta_\alpha(w/2K) \qquad (\alpha = 0, 1, 2, 3);$$

then

$$\operatorname{sn} w = \theta_3\theta_1(w)/\theta_2\theta_0(w).$$

Let us now turn to the functions $1 - x^2$ and $1 - k^2 x^2$. Recalling formula (6) in §24, we have that

$$\begin{aligned}
1 - x^2 &= 1 - \frac{e_1 - e_3}{\wp(w/\sqrt{e_1 - e_3}) - e_3} \\
&= \frac{\wp(w/\sqrt{e_1 - e_3}) - e_1}{\wp(w/\sqrt{e_1 - e_3}) - e_3} \\
&= \left\{ \frac{\sigma_1(w/\sqrt{e_1 - e_3})}{\sigma_3(w/\sqrt{e_1 - e_3})} \right\}^2 \\
&= \frac{\vartheta_0^2}{\vartheta_3^2} \left\{ \frac{\vartheta_2(w/2\omega\sqrt{e_1 - e_3})}{\vartheta_0(w/2\omega\sqrt{e_1 - e_3})} \right\}^2,
\end{aligned}$$

and hence

$$1 - x^2 = \frac{\theta_0^2}{\theta_2^2} \left\{ \frac{\theta_2(w)}{\theta_0(w)} \right\}^2,$$

from which

$$\sqrt{1 - x^2} = \operatorname{cn} w = \frac{\theta_0}{\theta_2}\frac{\theta_2(w)}{\theta_0(w)}.$$

Similarly,

$$\operatorname{dn} w = \frac{\theta_0}{\theta_3}\frac{\theta_3(w)}{\theta_0(w)}.$$

Primitive periods of these functions are given by the table

$\operatorname{cn} w$	$4K$	$2K + 2iK'$
$\operatorname{dn} w$	$2K$	$4iK'$

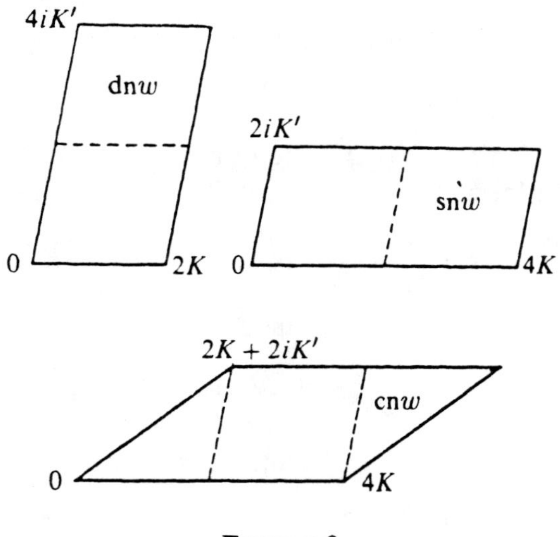

FIGURE 9

Period parallelograms for all three functions are represented in Figure 9.

We see that the parallelograms are all different, but they have the same area.

It is not hard to determine the zeros and poles of the Jacobi functions, as well as their values at certain other points. These facts are contained in Tables XIII and XV.

We underscore that the poles of the Jacobi functions are simple. Thus, we have here functions of the second type in the classification of §4.

The proof of the identity([16])

$$\mathrm{dn}^2(u; k)/\mathrm{cn}^2(u; k) = \mathrm{dn}^2(iu; k'),\tag{1}$$

which we need soon, is left as an exercise.

For a proof it suffices to verify that the left-hand and right-hand sides have the same primitive periods $2K$ and $2iK'$, have the same zeros and poles, and, finally, take the same value at the point $u = 0$.

In the Weierstrass theory arbitrary periods 2ω and $2\omega'$ could be specified. It was required only that the ratio $\tau = \omega'/\omega$ have nonzero (usually positive) imaginary part.

The situation is different here. The periods $2K$ and $2iK'$ cannot be chosen arbitrarily. Only the ratio $\tau = iK'/K$ or, what is the same, the

---

([16])This identity is a special case of certain relations that will be analyzed below with the necessary thoroughness, and that are contained in Table XIX.

quantity $h = e^{\pi i \tau} = e^{-\pi K'/K}$ can be specified arbitrarily. After this the periods are already determined, and we have the following formula for K:

$$K = \omega\sqrt{e_1 - e_3} = \frac{\pi}{2}\vartheta_3^2 = \frac{\pi}{2}(1 + 2h + 2h^4 + \cdots)^2.$$

On the other hand, a similar expression in terms of h can be written also for k^2. It has the form

$$k^2 = \frac{e_2 - e_3}{e_1 - e_3} = \frac{\vartheta_2^4}{\vartheta_3^4} = \left\{\frac{2h^{1/4} + 2h^{9/4} + \cdots}{1 + 2h + 2h^4 + \cdots}\right\}^4 \qquad (h^{1/4} = e^{\pi i\tau/4}).$$

In concluding this section we give Jacobi's original notation for the theta functions:

$$\begin{aligned} H(w) = \theta_1(w), & \qquad \Theta(w) = \theta_0(w),\\ H_1(w) = \theta_2(w), & \qquad \Theta_1(w) = \theta_3(w). \end{aligned}$$

This notation is also used at present, along with the notation above.

§26. Differentiation of the Jacobi functions

We take the integral

$$w = \int_0^x \frac{dt}{\sqrt{(1 - t^2)(1 - k^2 t^2)}}, \tag{1}$$

whose inverse is the function

$$x = \operatorname{sn} w. \tag{2}$$

It follows from (1) that

$$dx/dw = \sqrt{(1 - x^2)(1 - k^2 x^2)},$$

and therefore, by (2),

$$d\operatorname{sn} w/dw = \operatorname{cn} w \operatorname{dn} w.$$

To get the derivative of $\operatorname{cn} w$ and of $\operatorname{dn} w$ it is necessary to differentiate the relations

$$\operatorname{sn}^2 w + \operatorname{cn}^2 w = 1, \qquad k^2 \operatorname{sn}^2 w + \operatorname{dn}^2 w = 1,$$

which gives us that

$$d\operatorname{dn} w/dw = -\operatorname{sn} w \operatorname{dn} w,$$

$$d\operatorname{dn} w/dw = -k^2 \operatorname{sn} w \operatorname{cn} w.$$

We remark that the numbers $2K$ and $2iK'$ are primitive periods of the function $\operatorname{sn}^2 w$. This follows (without any investigation of the function $\operatorname{sn} w$) from formula (6) in §24, which can be written in the form

$$\operatorname{sn}^2 w = \frac{e_1 - e_3}{\wp(w/\sqrt{e_1 - e_3}) - e_3}. \tag{3}$$

As follows from the general theorems proved above, every elliptic funtion with periods $2K$ and $2iK'$ admits a representation

$$R_1(\wp) + R_2(\wp)\wp',$$

where

$$\wp = \wp(w/\sqrt{e_1 - e_3}).$$

Formula (3) now shows that every elliptic function with periods $2K$ and $2iK'$ admits a representation

$$R_1(\operatorname{sn}^2 w) + \operatorname{sn} w \operatorname{cn} w \operatorname{dn} w R_2(\operatorname{sn}^2 w).$$

This now gives us that an even elliptic function with periods $2K$ and $2iK'$ is equal to $R(\operatorname{sn}^2 w)$, and an odd elliptic function is equal to $\operatorname{sn} w \operatorname{cn} w \operatorname{dn} w R(\operatorname{sn}^2 w)$.

§27. The Jacobi function $Z(w)$

This function is analogous to the Weierstrass function $\zeta(u)$ and is defined by

$$Z(w) = \theta_0'(w)/\theta_0(w).$$

This is an odd function with one first-order pole $w = iK'$ in the period parallelogram. Since

$$\theta_0(w + 2K) = \theta_0(w),$$
$$\theta_0(w + 2iK') = -h^{-1}e^{-\pi i w/K}\theta_0(w),$$

it follows that

$$Z(w + 2K) = Z(w),$$
$$Z(w + 2iK') = Z(w) - \pi i/K.$$

Using the expansion of $\theta_0(w)$ in an infinite product, we get an expansion of $Z(w)$ in an infinite series. It has the form

$$Z(w) = \frac{2\pi}{K} \sum_{n=1}^{\infty} \frac{h^{2n-1}\sin(\pi w/K)}{1 - 2h^{2n-1}\cos(\pi w/K) + h^{4n-2}}.$$

As in the case of $\zeta(u)$, the function $Z(w)$ can be used for representing an arbitrary elliptic function.

Let us take the function

$$-k^2 \operatorname{sn} u \operatorname{sn} v \operatorname{sn}(u + v)$$

of u as an example. The numbers $2K$ and $2iK'$ are primitive periods of it. In the period parallelogram it has the poles $u = iK'$ and $u = -v + iK'$, both simple. Further, the residues are equal to -1 and 1, respectively. Therefore,

$$-k^2 \operatorname{sn} u \operatorname{sn} v \operatorname{sn}(u+v) = Z(u+v) - Z(u) + C.$$

To determine the constant C we set $u = 0$. This gives us that $Z(v) + C = 0$. Accordingly,

$$Z(u+v) - Z(u) - Z(v) = -k^2 \operatorname{sn} u \operatorname{sn} v \operatorname{sn}(u+v). \tag{1}$$

Similarly, it is not hard to get the following expansion into "simple fractions":

$$-\frac{2k^2 \operatorname{sn}^2 v \operatorname{sn} u \operatorname{cn} u \operatorname{dn} u}{1 - k^2 \operatorname{sn}^2 u \operatorname{sn}^2 v}$$
$$= Z(u+v) + Z(u-v) - 2Z(u). \tag{2}$$

However, this expansion can also be obtained from the expression for the function $1 - k^2 \operatorname{sn}^2 u \operatorname{sn}^2 v$ in terms of theta functions from the zeros and poles. The indicated expression has the form

$$1 - k^2 \operatorname{sn}^2 u \operatorname{sn}^2 v = \theta_0^2 \frac{\theta_0(u+v)\theta_0(u-v)}{\theta_0^2(u)\theta_0^2(v)},$$

and (2) is obtained from this by taking the logarithmic derivative of both sides. Interchanging u and v in (2) gives us that

$$-\frac{2k^2 \operatorname{sn}^2 u \operatorname{sn} v \operatorname{cn} v \operatorname{dn} v}{1 - k^2 \operatorname{sn}^2 u \operatorname{sn}^2 v}$$
$$= Z(u+v) - Z(u-v) - 2Z(v). \tag{2^{bis}}$$

We now add (2) and (2bis) and represent the result as

$$Z(u+v) - Z(u) - Z(v)$$
$$= -k^2 \operatorname{sn} u \operatorname{sn} v \frac{\operatorname{sn} u \operatorname{cn} v \operatorname{dn} v + \operatorname{sn} v \operatorname{cn} u \operatorname{dn} u}{1 - k^2 \operatorname{sn}^2 u \operatorname{sn}^2 v}. \tag{3}$$

Comparing this with (1), we get

$$\operatorname{sn}(u+v) = \frac{\operatorname{sn} u \operatorname{cn} v \operatorname{dn} v + \operatorname{sn} v \operatorname{cn} u \operatorname{dn} u}{1 - k^2 \operatorname{sn}^2 u \operatorname{sn}^2 v}. \tag{4}$$

This relation expresses the addition theorem for the function $\operatorname{sn} u$.

EXERCISE. Prove that

$$\operatorname{sn}^2 \frac{K}{2} = \frac{1}{1 + \sqrt{1 - k^2}}.$$

§28. The Euler theorem

Addition theorems for other functions can be obtained from the addition theorem (4) derived in §27 for $\operatorname{sn} u$. For example, to get the addition theorem for cn u we can start from the fact that cn $u = \sqrt{1 - \text{sn}^2 u}$. We do not give the computations, which are neither difficult nor interesting, but refer the reader to Table XIV, which contains more important formulas expressing addition theorems.

Here we dwell on another aspect of the question. The fact of the matter is that formula (4) in §27 can be obtained on the basis of considerations going back to Euler and not having anything in common with the theory we have constructed.

Euler considered the differential equation

$$\frac{dx}{\sqrt{f(x)}} + \frac{dy}{\sqrt{f(y)}} = 0,$$

where $f(z) = a_0 z^4 + 4a_1 z^3 + 6a_2 z^2 + 4a_3 z + a_4$, and showed that this equation has an algebraic integral. Comparison of this integral with the transcendental integral

$$\int \frac{dx}{\sqrt{f(x)}} + \int \frac{dy}{\sqrt{f(y)}} = C$$

leads to the addition theorem for the Jacobi functions. Generally speaking, this theorem of Euler served as the first impetus for the investigation of elliptic integrals.

An especially elegant method of proving the Euler theorem is due to Darboux. Following him, let us consider the equation

$$\frac{dx}{\sqrt{(1 - x^2)(1 - k^2 x^2)}} + \frac{dy}{\sqrt{(1 - y^2)(1 - k^2 y^2)}} = 0. \tag{1}$$

It has the transcendental integral

$$\int_0^x \frac{dx}{\sqrt{(1 - x^2)(1 - k^2 x^2)}} + \int_0^y \frac{dy}{\sqrt{(1 - y^2)(1 - k^2 y^2)}} = A, \tag{2}$$

where A is an arbitrary constant. If we let

$$u = \int_0^x \frac{dx}{\sqrt{(1 - x^2)(1 - k^2 x^2)}}, \tag{3_1}$$

$$v = \int_0^y \frac{dy}{\sqrt{(1 - y^2)(1 - k^2 y^2)}}, \tag{3_2}$$

then the integral (2) can be represented in the form

$$u + v = A. \tag{2^{bis}}$$

On the other hand, equation (1) can be replaced by the system

$$\left.\begin{array}{l} dx/dt = \sqrt{(1 - x^2)(1 - k^2 x^2)}, \\ dy/dt = -\sqrt{(1 - y^2)(1 - k^2 y^2)}. \end{array}\right\} \tag{4}$$

Squaring (4), we get

$$\left.\begin{array}{l}(dx/dt)^2 = (1-x^2)(1-k^2x^2), \\ (dy/dt)^2 = (1-y^2)(1-k^2y^2).\end{array}\right\} \qquad (5)$$

Let us now differentiate these equations:

$$d^2x/dt^2 = x(2k^2x^2 - 1 - k^2), \qquad d^2y/dt^2 = y(2k^2y^2 - 1 - k^2).$$

From this we get

$$y\frac{d^2x}{dt^2} - x\frac{d^2y}{dt^2} = 2k^2xy(x^2 - y^2),$$

or

$$\frac{d}{dt}\left(y\frac{dx}{dt} - x\frac{dy}{dt}\right) = 2k^2xy(x^2 - y^2). \qquad (6)$$

On the other hand, it follows from (5) that

$$y^2\left(\frac{dx}{dt}\right)^2 - x^2\left(\frac{dy}{dt}\right)^2 = (y^2 - x^2)(1 - k^2x^2y^2). \qquad (7)$$

Dividing (6) by (7), we get

$$\frac{\frac{d}{dt}(y\frac{dx}{dt} - x\frac{dy}{dt})}{y\frac{dx}{dt} - x\frac{dy}{dt}} = \frac{2k^2xy(y\frac{dx}{dt} + x\frac{dy}{dt})}{k^2x^2y^2 - 1},$$

or

$$\frac{d}{dt}\ln\left(y\frac{dx}{dt} - x\frac{dy}{dt}\right) = \frac{d}{dt}\ln(k^2x^2y^2 - 1),$$

from which

$$y\frac{dx}{dt} - x\frac{dy}{dt} = C(1 - k^2x^2y^2).$$

Taking (4) into account, we get

$$\frac{x\sqrt{(1-y^2)(1-k^2y^2)} + y\sqrt{(1-x^2)(1-k^2x^2)}}{1 - k^2x^2y^2} = C. \qquad (8)$$

This is the algebraic form of the integral of equation (1).

Each of the relations (2) and (8) must be a consequence of the other. The addition theorem for $\operatorname{sn} w$ is obtained from this.

Indeed, by (3_1) and (3_2) we have $x = \operatorname{sn} u$ and $y = \operatorname{sn} v$. Therefore, (8) can be represented in the form

$$\frac{\operatorname{sn} u \operatorname{cn} v \operatorname{dn} v + \operatorname{sn} v \operatorname{cn} u \operatorname{dn} u}{1 - k^2 \operatorname{sn}^2 u \operatorname{sn}^2 v} = C. \qquad (8^{\text{bis}})$$

Since (8bis) is a consequence of (2bis), C is a function of $A: C = \varphi(A)$. In other words,

$$\frac{\operatorname{sn} u \operatorname{cn} v \operatorname{dn} v + \operatorname{sn} v \operatorname{cn} u \operatorname{dn} u}{1 - k^2 \operatorname{sn}^2 u \operatorname{sn}^2 v} = \varphi(u + v).$$

To determine the form of φ we set $v = 0$. This gives us that $\operatorname{sn} u = \varphi(u)$, and hence

$$\frac{\operatorname{sn} u \operatorname{cn} v \operatorname{dn} v + \operatorname{sn} v \operatorname{cn} u \operatorname{dn} u}{1 - k^2 \operatorname{sn}^2 u \operatorname{sn}^2 v} = \operatorname{sn}(u + v).$$

§29. Normal elliptic integrals of the second and third kinds in Jacobi form

We recall the contents of §17, where it was shown that every elliptic integral in Weierstrass form can be expressed in terms of an elliptic function and the following three integrals:

$$u = \int \frac{dz}{w}, \tag{1}$$

$$\zeta(u) = -\int \frac{z \, dz}{w}, \tag{2}$$

$$\ln \frac{\sigma(u - u_0)}{\sigma(u)} + u\zeta(u_0) = \frac{1}{2} \int \frac{w + w_0}{z - z_0} \frac{dz}{w}, \tag{3}$$

where $w^2 = 4z^3 - g_2 z - g_3$, $z_0 = \wp(u_0)$, and $w_0 = \wp'(u_0)$.

We now want to consider elliptic integrals in the Jacobi and Riemann forms. To transform the basic integrals (1), (2), and (3) to Riemann form it is necessary to replace w^2 by $4z(1 - z)(1 - k^2 z)$ in them. If we then make the substitution $z = t^2$, we arrive at the following integrals:

$$\int \frac{dt}{\sqrt{(1 - t^2)(1 - k^2 t^2)}}, \tag{1^{bis}}$$

$$\int \frac{t^2 \, dt}{\sqrt{(1 - t^2)(1 - k^2 t^2)}}, \tag{2^{bis}}$$

$$\int \frac{t\sqrt{(1 - t^2)(1 - k^2 t^2)} + t_0\sqrt{(1 - t_0^2)(1 - k^2 t_0^2)}}{(t^2 - t_0^2)\sqrt{(1 - t^2)(1 - k^2 t^2)}} \, dt. \tag{3^{bis}}$$

The integral (1bis), which we rewrite in the form

$$u = \int_0^x \frac{dt}{\sqrt{(1 - t^2)(1 - k^2 t^2)}}, \tag{4}$$

is an integral of the first kind in the Jacobi theory.

The integral

$$\int_0^x \frac{(1 - k^2 t^2) \, dt}{\sqrt{(1 - t^2)(1 - k^2 t^2)}} \tag{$5'$}$$

is taken as a normal integral of the second kind instead of (2bis) in the Jacobi theory. This expression differs from (2bis) by an integral of the first kind (if we do not consider a constant factor).

The integral $(5')$ can be expressed as a function of the integral of the first kind (4). The corresponding notation of Jacobi is

$$E(u) = \int_0^x \sqrt{\frac{1 - k^2 t^2}{1 - t^2}}\, dt \qquad [x = \mathrm{sn}(u; k)], \tag{5}$$

or

$$E(u) = \int_0^u \mathrm{dn}^2(v; k)\, dv.$$

Finally, instead of (3bis) Jacobi takes

$$\int_0^x \frac{k^2 b \sqrt{(1 - b^2)(1 - k^2 b^2)}}{1 - k^2 b^2 t^2} \frac{t^2\, dt}{\sqrt{(1 - t^2)(1 - k^2 t^2)}} \tag{6'}$$

as a normal integral of the third kind, and this differs from (3bis) by an elementary function and an integral of the first kind (if we do not consider an additional constant factor).

The integral $(6')$ can be expressed as a function of the integral of the first kind (4). Jacobi denotes it by $\Pi(u; a)$ if $b = \mathrm{sn}(a; k)$. Thus,

$$\begin{aligned}
\Pi(u; a) &= \int_0^x \frac{k^2 b \sqrt{(1 - b^2)(1 - k^2 b^2)}}{1 - k^2 b^2 t^2} \frac{t^2\, dt}{\sqrt{(1 - t^2)(1 - k^2 t^2)}} \\
&= \int_0^u \frac{k^2 \,\mathrm{sn}\, a \,\mathrm{cn}\, a \,\mathrm{dn}\, a \,\mathrm{sn}^2 v}{1 - k^2 \,\mathrm{sn}^2 a \,\mathrm{sn}^2 v}\, dv.
\end{aligned} \tag{6}$$

It was shown in §27, (2bis), that

$$\frac{k^2 \,\mathrm{sn}\, a \,\mathrm{cn}\, a \,\mathrm{dn}\, a \,\mathrm{sn}^2 v}{1 - k^2 \,\mathrm{sn}^2 a \,\mathrm{sn}^2 v} = \frac{1}{2} Z(v - a) - \frac{1}{2} Z(v + a) + Z(a).$$

But since $Z(v) = \Theta'(v)/\Theta(v)$, it follows that

$$\Pi(u, a) = \frac{1}{2} \ln \frac{\Theta(u - a)}{\Theta(u + a)} + u Z(a).$$

This function in the Jacobi theory replaces the function

$$\ln \frac{\sigma(u - a)}{\sigma(u)} + u \zeta(a)$$

in the Weierstrass theory.

It is not hard to express the normal integral of the second kind $E(u)$ in terms of the functions introduced earlier. With this goal we again take formula (2bis) in §27:

$$Z(u + v) - Z(u - v) - 2Z(v) = -\frac{2k^2 \,\mathrm{sn}^2 u \,\mathrm{sn}\, v \,\mathrm{cn}\, v \,\mathrm{dn}\, v}{1 - k^2 \,\mathrm{sn}^2 u \,\mathrm{sn}^2 v}.$$

Let us divide both sides by $2 \operatorname{sn} v$ and let v go to 0. In the limit we obtain

$$\frac{d}{du} Z(u) - Z'(0) = -k^2 \operatorname{sn}^2 u,$$

from which

$$\operatorname{dn}^2 u = 1 - Z'(0) + \frac{d}{du} Z(u),$$

and hence

$$E(u) = [1 - Z'(0)]u + Z(u).$$

We do not dwell here on another representation of the first term on the right-hand side (see Table XVI, and also §31).

§30. Complete elliptic integrals of the first kind

In §25 we introduced the quantities K and K' as the functions of τ or $h = e^{\pi i \tau}$ ($\Im \tau > 0$) defined by the following formulas:

$$K = \tfrac{\pi}{2} \vartheta_3^2(0|\tau) = \tfrac{\pi}{2}(1 + 2h + 2h^4 + \cdots)^2, \qquad iK'/K = \tau. \qquad (1)$$

Moreover, we had the formula

$$k^2 = \frac{\vartheta_2^4(0|\tau)}{\vartheta_3^4(0|\tau)} = \left\{ \frac{2h^{1/4} + 2h^{9/4} + \cdots}{1 + 2h + 2h^4 + \cdots} \right\}^4 \qquad (h^{1/4} = e^{\pi i \tau/4}), \qquad (2)$$

which implies that $k^2 = \lambda(\tau)$.

Let us now study K as a function of $k^2 = \lambda$. We recall the conformal mapping of D_2 generated by the function $\lambda(\tau)$. The values of the function K defined by (1) are the same at two points lying in the λ-plane on opposite edges of the cut BC (from $\lambda = -\infty$ to $\lambda = 0$), because to such points there correspond points on the rectilinear boundaries of D_2 that are symmetric with respect to the imaginary axis, and $h = e^{2\pi i \tau}$ has the same value at these points. Therefore, the quantity K defined by (1) is a regular function of $k^2 = \lambda$ in the λ-plane, cut along the single interval $(1, \infty)$ alone. We prove that

$$K = \int_0^1 \frac{dt}{\sqrt{(1 - t^2)(1 - k^2 t^2)}} \qquad (3)$$

in the plane cut in this way, where the radical is equal to 1 when $t = 0$; this will always be assumed.

Since both sides of (3) are analytic in the domain being considered, it suffices to prove that (3) is true for $0 < k^2 < 1$. To prove this we note that the only points u at which

$$\operatorname{sn} u = 1, \qquad (4)$$

are

$$u = (4m + 1)K + 2m'iK',$$

where m and m' are integers. Indeed, the function $\operatorname{sn} u$ is of second order, and the root $u = K$ of (4) is a double root, since the derivative

$$d\operatorname{sn} u/du = \operatorname{cn} u\, \operatorname{dn} u$$

vanishes at $u = K$. Thus, for some integers m and m'

$$\int_0^1 \frac{dt}{\sqrt{(1 - t^2)(1 - k^2 t^2)}} = (4m + 1)K + 2m'iK'.$$

Let $0 < k^2 < 1$. Then τ is purely imaginary, and K and K' are positive. Therefore, $m' = 0$,

$$\int_0^1 \frac{dt}{\sqrt{(1 - t^2)(1 - k^2 t^2)}} = (4m + 1)K > 0,$$

and hence $m > -1$. However, m cannot be greater than 0, because then for some positive $x < 1$ we would have

$$\int_0^x \frac{dt}{\sqrt{(1 - t^2)(1 - k^2 t^2)}} = K,$$

which would imply that $0 < x = \operatorname{sn} K < 1$.

Hence, $m = 0$, and our assertion is proved, i.e., the quantity K determined by (1) is representable in the form (3) if k^2 does not belong to $[1, \infty)$.

Making the substitution $t = \sin \psi$ in (3), we get that

$$\int_0^{\pi/2} \frac{d\psi}{\sqrt{1 - k^2 \sin^2 \psi}} = K.$$

As we mentioned earlier, Legendre regarded the integral

$$\int_0^{\varphi} \frac{d\psi}{\sqrt{1 - k^2 \sin^2 \psi}} = F(\varphi, k)$$

as a function of the angle φ and the modulus k. The angle φ varied in the interval $[0, \pi/2]$. The quantity K is obtained for $\varphi = \pi/2$:

$$K = F(\pi/2, k).$$

Therefore, Legendre called this quantity the *complete* elliptic integral of the first kind for the modulus k.

Along with the modulus k one often has to consider the so-called complementary modulus k' determined by the formula $k^2 + k'^2 = 1$. We prove that

$$\int_0^1 \frac{dt}{\sqrt{(1 - t^2)(1 - k'^2 t^2)}} = K'. \tag{5}$$

Here it must be assumed that k'^2 does not lie on the real semi-axis from $\lambda = 1$ to $\lambda = \infty$. This means that k^2 must not lie on the real semi-axis from $\lambda = 0$ to $\lambda = -\infty$. Thus, when both quantities K and K' are considered at the same time it must be assumed that the λ-plane is cut along the two semi-axes from $\lambda = 1$ to $\lambda = \infty$ and from $\lambda = 0$ to $\lambda = -\infty$.

To prove (5) we take into account the relation

$$\vartheta_3(0|\tau) = (-i\tau)^{-1/2}\vartheta_3(0|-1/\tau),$$

from which

$$\vartheta_3^2(0|\tau) = \tfrac{i}{\tau}\vartheta_3^2(0|-1/\tau).$$

But since

$$K' = \tfrac{1}{i}\tau K = \tfrac{\tau}{i}\tfrac{\pi}{2}\vartheta_3^2(0|\tau),$$

it follows that

$$K' = \tfrac{\pi}{2}\vartheta_3^2(0|-1/\tau). \tag{1'}$$

On the other hand,

$$1 - k^2 = \frac{\vartheta_0^4(0|\tau)}{\vartheta_3^4(0|\tau)} = \frac{\vartheta_2^4(0|-1/\tau)}{\vartheta_3^4(0|-1/\tau)},$$

i.e.,

$$k'^2 = \frac{\vartheta_2^4(0|-1/\tau)}{\vartheta_3^4(0|-1/\tau)}. \tag{2'}$$

Comparison of the pair of formulas (1′), (2′) with the pair (1), (2) shows that K' depends on k'^2 as K depends on k^2, and this proves our assertion.

To conclude this section, we make a remark on the passage from the Weierstrass form to the Jacobi form.

We have seen that k^2 can be expressed in a definite way in terms of the roots e_α of the polynomial $4x^3 - g_2 x - g_3$, namely,

$$k^2 = (e_2 - e_3)/(e_1 - e_3).$$

If we want k^2 not to take values lying on the real semi-axes from $\lambda = 1$ to $\lambda = \infty$ and from $\lambda = 0$ to $\lambda = -\infty$, then the numbering of the roots must be subject to the following condition: if e_1, e_2, and e_3 lie on a single line, then e_2 must lie between e_1 and e_3.

We mention also that the case when $0 < k^2 < 1$ is the most important for applications. This case occurs if all the roots e_α are real. It is then usually assumed that $e_1 > e_2 > e_3$. This case is sometimes called the *normal case*.

§31. Complete elliptic integrals of the second kind

These integrals for the modulus k and the complementary modulus k' are defined by

$$E = \int_0^1 \sqrt{\frac{1 - k^2 t^2}{1 - t^2}}\, dt, \qquad E' = \int_0^1 \sqrt{\frac{1 - k'^2 t^2}{1 - t^2}}\, dt.$$

It is easy to see that $E = E(K)$, where $E(u)$ denotes the integral of the second kind considered in §29. It was shown in the same place that

$$E(u) = [1 - Z'(0)]u + Z(u).$$

Since $Z(K) = 0$, it follows that $E = [1 - Z'(0)]K$, and hence

$$E(u) = Z(u) + (E/K)u. \tag{1}$$

It was proved in §12 that $\eta\omega' - \eta'\omega = \pi i/2$ if $\Im\tau > 0$. Some relation in the Jacobi theory must correspond to this relation in the Weierstrass theory, and the quantities K, K', E, and E' must appear in it instead of ω, ω', η, and η'.

This relation, which is due to Legendre and bears his name, has the form

$$EK' + E'K - KK' = \pi/2.$$

To prove the Legendre relation we derive some transformation formulas for $Z(u)$ and $E(u)$ connected with the transition from τ to $\tau' = -1/\tau$. Therefore, instead of $Z(u)$ and $E(u)$ we write $Z(u; k)$ and $E(u; k)$, recalling that the transition from k to k' corresponds to the transition from τ to τ'.

It was proved in §22 that

$$\vartheta_0(\tau' v | \tau') = \sqrt{-i\tau}\,\vartheta_2(v | \tau)e^{-\pi i \tau' v^2}.$$

Replacing v by $u/2K$, we get

$$\vartheta_0\left(\frac{iu}{2K'}\,\Big|\,\tau'\right) = \sqrt{-i\tau}\,\vartheta_2\left(\frac{u}{2K}\,\Big|\,\tau\right)e^{\pi u^2/4KK'},$$

or

$$\Theta(iu; k') = \sqrt{-i\tau}\,H_1(u; k)e^{\pi u^2/4KK'}.$$

The quantity $H_1(u; k)$ is equal to

$$\sqrt{k/k'}\,\operatorname{cn}(u; k)\Theta(u; k).$$

Therefore,

$$\Theta(iu; k') = C\operatorname{cn}(u; k)\Theta(u; k)e^{\pi u^2/4KK'}.$$

From this, taking the logarithmic derivative with respect to u, we get

$$iZ(iu; k') = -\frac{\operatorname{sn}(u; k)\,\operatorname{dn}(u; k)}{\operatorname{cn}(u; k)} + Z(u; k) + \frac{\pi u}{2KK'}. \tag{2}$$

Observe now that

$$\frac{d}{du}\frac{\operatorname{sn}(u; k)\,\operatorname{dn}(u; k)}{\operatorname{cn}(u; k)} = -1 + \operatorname{dn}^2(u; k) + \frac{\operatorname{dn}^2(u; k)}{\operatorname{cn}^2(u; k)}.$$

Therefore, it follows from the identity (1) in §25 that

$$\frac{d}{du}\frac{\operatorname{sn}(u; k)\,\operatorname{dn}(u; k)}{\operatorname{cn}(u; k)} = -1 + \operatorname{dn}^2(u; k) + \operatorname{dn}^2(iu; k').$$

But since

$$E(u; k) = \int_0^u \operatorname{dn}^2(u; k)\, du,$$

$$E(iu; k') = i \int_0^u \operatorname{dn}^2(u; k')\, du,$$

it follows that

$$\frac{\operatorname{sn}(u; k)\,\operatorname{dn}(u; k)}{\operatorname{cn}(u; k)} = -u + E(u; k) - iE(iu; k'),$$

and hence (2) can be represented in the form

$$i\{Z(iu; k') - E(iu; k')\} = u(1 + \pi/2KK') + \{Z(u; k) - E(u; k)\}. \tag{3}$$

By (1), it follows from (3) that

$$-\frac{E'}{K'}u = \frac{E}{K}u - u\left(1 + \frac{\pi}{2KK'}\right);$$

from this,

$$-E'K = EK' - KK' - \pi/2,$$

which represents the Legendre formula, and the latter is thus proved.[17]

In concluding this section we mention the relations

$$\left.\begin{array}{l} E(u + 2K) = E(u) + 2E, \\ E(u + 2iK') = E(u) + 2i(K' - E'). \end{array}\right\} \tag{4}$$

Here $E(u) = E(u; k)$. It follows from these relations that E and $i(K' - E')$ play the same role in the Jacobi theory that η and η' play in the Weierstrass theory. The first relation in (4) follows from (1) by the fact that

$$Z(u + 2K) = Z(u).$$

To obtain the second relation in (4) it is necessary to take the equalities

$$Z(u + 2iK') = Z(u) + Z(2iK'), \qquad Z(2iK') = -\pi i/K.$$

They imply that

$$E(u + 2iK') = E(u) + 2i(EK' - \pi/2)/K,$$

---

[17] Another proof is given in §44.

and it remains to consider that, by the Legendre relation,

$$(EK' - \pi/2)/K = K' - E'.$$

§32. Degeneracy of elliptic functions

There is degeneracy in the elementary functions when one or both periods become infinitely large. In the first case the trigonometric functions are degenerate functions, and in the second case the rational functions are degenerate.

Assume that ω remains finite, while ω' tends to infinity. From the infinite product defining the sigma function it is easy to see that for this degeneracy the function $\sigma(u)$ becomes

$$\frac{2\omega}{\pi} e^{1/3!(\pi u/2\omega)^2} \sin \frac{\pi u}{2\omega}. \tag{1}$$

Similarly, from the series defining $\zeta(u)$ and $\wp(u)$ it is easy to see that $\zeta(u)$ degenerates into

$$\frac{1}{3}\left(\frac{\pi}{2\omega}\right)^2 u + \frac{\pi}{2\omega} \cot \frac{\pi u}{2\omega}, \tag{2}$$

and $\wp(u)$ degenerates into

$$-\frac{1}{3}\left(\frac{\pi}{2\omega}\right)^2 + \left(\frac{\pi}{2\omega}\right)^2 \frac{1}{\sin^2(\pi u/2\omega)}. \tag{3}$$

Further formulas relating to this case are contained in Table VII.

If the second period also tends to infinity, then instead of (1)–(3) we arrive at

$$u, \tag{1'}$$

$$1/u, \tag{2'}$$

$$1/u^2, \tag{3'}$$

which can be obtained both from (1)–(3) and directly from the definitions of $\sigma(u)$, $\zeta(u)$, and $\wp(u)$ by means of an infinite product and infinite series.

Let us now proceed to h, K, K', E, and E' as functions of k^2, and investigate how they behave as $k \to 0$.

From the equality

$$K = \int_0^1 \frac{dt}{\sqrt{(1 - t^2)(1 - k^2 t^2)}}$$

it follows that

$$\lim_{k \to 0} K = \pi/2.$$

Similarly,

$$\lim_{k\to 0} E = \pi/2 \quad \text{and} \quad \lim_{k\to 0} E' = 1.$$

We turn now to the formula

$$k^2 = \left(\frac{2h^{1/4} + 2h^{9/4} + \cdots}{1 + 2h + 2h^4 + \cdots}\right)^4.$$

It implies that

$$\lim_{k\to 0} \frac{h}{k^2} = \frac{1}{16},$$

and, on the other hand,

$$K' = \frac{\tau}{i}K = -\frac{K}{\pi}\ln h;$$

therefore,

$$K' - \frac{2K}{\pi}\ln\frac{4}{k} = -\frac{K}{\pi}\ln\frac{16h}{k^2}.$$

The right-hand side tends to 0 as $k \to 0$. Consequently,

$$\lim_{k\to 0}\left(K' - \frac{2K}{\pi}\ln\frac{4}{k}\right) = 0,$$

which now easily leads to the conclusion that

$$\lim_{k\to 0}\left(K' - \ln\frac{4}{k}\right) = 0.$$

The behavior of the quantities under consideration as $k^2 \to 1$ does not need to be investigated specially, since $k^2 \to 1$ as $k' \to 0$.

The degeneracy of the functions $\mathrm{sn}(u; k)$, $\mathrm{cn}(u; k)$, and $\mathrm{dn}(u; k)$ as $k \to 0$ and $k^2 \to 1$ is most simply investigated with the help of the integral relation

$$u = \int_0^x \frac{dt}{\sqrt{(1 - t^2)(1 - k^2 t^2)}},$$

which connects the function $x = \mathrm{sn}(u; k)$ with its argument u. From this it is clear that

$$\lim_{k\to 0} \mathrm{sn}(u; k) = \sin u.$$

Further, if

$$\xi = \lim_{k^2\to 1} \mathrm{sn}(u; k),$$

then

$$u = \int_0^\xi \frac{dt}{1 - t^2} = \frac{1}{2}\ln\frac{1 + \xi}{1 - \xi}.$$

Consequently,

$$\lim_{k^2 \to 1} \operatorname{sn}(u; k) = \frac{e^u - e^{-u}}{e^u + e^{-u}} = \frac{\operatorname{sh} u}{\operatorname{ch} u}.$$

§33. The simple pendulum

The most trivial way in which elliptic functions are used in mechanics and engineering is their introduction for integrating a differential equation or for computing an integral. Following the widespread tradition, we consider the problem of oscillation of a simple pendulum as a typical example of this kind of application.

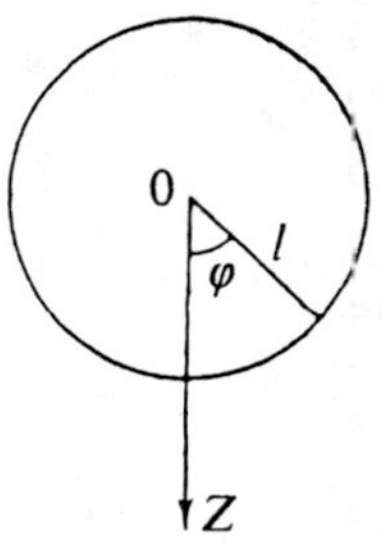

FIGURE 10

We take the point of suspension as the origin of coordinates, and direct the Z-axis vertically downward (Figure 10). Let the length of the pendulum be l, and let the initial position be its lowest position ($z = l$). The speed of the pendulum is denoted by v, and the initial speed by v_0 (> 0). Denoting by g the acceleration of gravity, we have the kinetic energy integral

$$v^2/2 - gz = -ga. \tag{1}$$

Since $v = v_0$ for $z = l$, the constant a is equal to

$$a = l - v_0^2/2g.$$

As an unknown function we introduce instead of z the angle φ between the Z-axis and the pendulum. Then $z = l \cos \varphi$ and $v = l \, d\varphi/dt$. Consequently, the kinetic energy equation can be written in the form

$$\frac{l^2}{2} \left(\frac{d\varphi}{dt} \right)^2 - gl \cos \varphi = -ga,$$

or

$$\frac{l \, d\varphi}{\sqrt{2g(l \cos \varphi - a)}} = dt,$$

which implies that

$$t = \frac{l}{\sqrt{2g}} \int_0^\varphi \frac{d\psi}{\sqrt{l\cos\psi - a}}. \tag{2}$$

We analyze three cases, depending on whether the constant a is greater than, equal to, or less than $-l$.

First case: $a > -l$. This case holds if $v_0 < 2\sqrt{gl}$, i.e., if the initial speed v_0 is not "very" large. We introduce the angle α ($0 < \alpha < \pi$) by means of the relation $a = l\cos\alpha$. Then (2) takes the form

$$t = \sqrt{\frac{l}{2g}} \int_0^\varphi \frac{d\psi}{\sqrt{\cos\psi - \cos\alpha}},$$

or

$$t = \frac{1}{2}\sqrt{\frac{l}{g}} \int_0^\varphi \frac{d\psi}{\sqrt{\sin^2(\alpha/2) - \sin^2(\psi/2)}}.$$

Setting $k^2 = \sin^2(\alpha/2)$ and

$$\sin(\psi/2) = x\sin(\alpha/2), \qquad \sin(\varphi/2) = u\sin(\alpha/2),$$

we get that

$$t = \sqrt{\frac{l}{g}} \int_0^u \frac{dx}{\sqrt{(1-x^2)(1-k^2x^2)}},$$

which implies that

$$u = \operatorname{sn}\left(\sqrt{g/l}\,t; k\right).$$

Consequently,

$$\sin(\varphi/2) = \sin(\alpha/2)\operatorname{sn}(\sqrt{g/l}\,t; \sin(\alpha/2)).$$

This formula expresses the law of motion of the pendulum, and easily allows us to see all the features of this motion.

First, the motion is periodic and the period is equal to $4\sqrt{l/g}K$. Second, occupying the lowest position at $t = 0$, the pendulum occupies the highest position ($\varphi = \alpha$) after a quarter-period. After another quarter-period it again occupies the lowest position. Finally, after a further quarter-period the pendulum will occupy the highest position on the other side of the vertical axis ($\varphi = -\alpha$). Third, the speed of the pendulum is equal to zero at the highest position.

Second case: $a = -l$. This is a limit case for the preceding case and occurs when $v_0 = 2\sqrt{gl}$. Since now $\alpha = \pi$, we have the equations

$$\sin\frac{\varphi}{2} = u \quad \text{and} \quad t = \sqrt{\frac{l}{g}} \int_0^u \frac{dx}{1-x^2}.$$

Integration gives us that

$$t = \frac{1}{2}\sqrt{\frac{l}{g}}\,\ln\frac{1+u}{1-u},$$

which implies that

$$\sin(\varphi/2) = \tanh\left(t\sqrt{g/l}\right).$$

This formula shows that as t increases from 0 to ∞ the angle φ increases monotonically from 0 to π, i.e., the pendulum always moves in one direction, and the uppermost position, which it never attains, is its limit position.

Third case: $a < -l$. In this case, which happens when $v_0 > 2\sqrt{gl}$, (2) can be rewritten in the form

$$t = \frac{l}{\sqrt{2g}}\int_0^\varphi \frac{d\psi}{\sqrt{l(1 - 2\sin^2(\psi/2)) - a}}. \tag{3}$$

We set $k^2 = 2l/(l - a)$, so that $0 < k^2 < 1$, and also $\sin(\psi/2) = x$ and $\sin(\varphi/2) = u$. Then (3) takes the form

$$t = \sqrt{\frac{2l}{l-a}}\sqrt{\frac{l}{g}}\int_0^u \frac{dx}{\sqrt{(1 - x^2)(1 - k^2 x^2)}}.$$

From this,

$$u = \operatorname{sn}\left(\sqrt{\frac{l-a}{2l}}\sqrt{\frac{g}{l}}\,t; k\right),$$

and hence

$$\sin\frac{\varphi}{2} = \operatorname{sn}\left(\sqrt{\frac{l-a}{2l}}\sqrt{\frac{g}{l}}\,t;\sqrt{\frac{2l}{l-a}}\right).$$

This formula shows that the pendulum attains the highest position $z = -l$ for t equal to

$$\frac{\sqrt{2l}}{\sqrt{g(l-a)}}K.$$

However, in this position its speed is nonzero. In general, the speed never becomes zero, since this could happen (as (1) shows) only for $z = a$; but $a < -l$, and z always lies in the interval $[-l, l]$. Thus, instead of an oscillatory motion the motion is now always in a circle in a single direction.

CHAPTER 6

Transformation of Elliptic Functions

§34. The problem of transformation of elliptic functions

This problem can be formulated as follows in one of its variants: find conditions under which the differential equation

$$dx/\sqrt{f(x)} = dy/\sqrt{g(y)}, \tag{1}$$

where $f(x)$ and $g(y)$ are polynomials of fourth or third degree, has an algebraic integral, i.e., an integral of the form $F(x,y) = 0$, where F is a polynomial in its arguments; and find this integral if the indicated conditions hold. In other words, we are concerned with transforming the elliptic differential

$$dx/\sqrt{f(x)} \tag{2}$$

into the elliptic differential

$$dy/\sqrt{g(y)} \tag{3}$$

by means of an algebraic relation $F(x,y) = 0$.

We have already dealt with certain special cases of this general problem.

First, it was shown in §17 that the differential (2) can always be reduced to the normal Weierstrass form

$$dy/\sqrt{4y^3 - g_2 y - g_3} \tag{2'}$$

by means of a suitable linear transformation

$$x = (\alpha y + \beta)/(\gamma y + \delta). \tag{4}$$

Thus, the differential equation

$$\frac{dx}{\sqrt{f(x)}} = \frac{dy}{\sqrt{4y^3 - g_2 y - g_3}},$$

which is a special case of (1), certainly admits a linear integral (4) if g_2 and g_3 are invariants of the polynomial $f(x)$.

87

Second, the Euler theorem was proved in §28, and in view of it the equation

$$dx/\sqrt{f(x)} = dy/\sqrt{f(y)}$$

has an algebraic integral. The Euler equation is the special case of (1) when $g(y) = f(y)$.

The importance of the general transformation problem can be judged already from these two special cases, of which the second expresses one of the basic properties of elliptic functions—the existence of an algebraic addition theorem—while the first allows us to confine ourselves to the investigation of certain standard forms of elliptic integrals, a circumstance that is especially useful in computations in general and when working with tables in particular.

The desire to reduce a given elliptic differential (2) to a simple form (3), although with the help of a complicated algebraic dependence between x and y, served perhaps as a stimulus for the development of the general theory of transformations of elliptic functions, which was created mainly by Abel and Jacobi.

To investigate the general transformation problem it is useful first to reduce it to certain simpler problems.

The first simplification we can make consists in replacing (1) by the equation

$$\frac{dx}{\sqrt{4x(1-x)(1-k^2x)}} = M\frac{dy}{\sqrt{4y(1-y)(1-\lambda^2y)}}, \tag{5}$$

where M is a constant.

Indeed, as shown in §24, the linear transformation

$$y = e_3 + (e_1 - e_3)/z$$

carries the differential (2′) into the differential

$$-\frac{1}{\sqrt{e_1 - e_3}}\frac{dz}{\sqrt{4z(1-z)(1-k^2z)}},$$

where

$$k^2 = (e_2 - e_3)/(e_1 - e_3).$$

If both sides of (1) are transformed to a similar form, equation (5) is obtained.

The second simplification consists in confining ourselves to the determination of the integral of (5) that assigns to the point $x = 0$ the point $y = 0$, in other words, the determination of the algebraic dependence between x and y that follows from the relation

$$\int_0^x \frac{dt}{\sqrt{4t(1-t)(1-k^2t)}} = M \int_0^y \frac{dt}{\sqrt{4t(1-t)(1-\lambda^2t)}}. \tag{6}$$

Indeed, the general relation between x and y has the form

$$\int_a^c \frac{dt}{\sqrt{4t(1-t)(1-k^2t)}} = M \int_0^y \frac{dt}{\sqrt{4t(1-t)(1-\lambda^2t)}}$$

where a is a constant. If we find the algebraic dependence between z and y that follows from the special relation

$$\int_0^z \frac{dt}{\sqrt{4t(1-t)(1-k^2t)}} = M \int_0^y \frac{dt}{\sqrt{4t(1-t)(1-\lambda^2t)}},$$

then everything reduces to the determination of the algebraic dependence between x and z that follows from the relation

$$\int_0^z \frac{dt}{\sqrt{4t(1-t)(1-k^2t)}}$$
$$= \int_0^x \frac{dt}{\sqrt{4t(1-t)(1-k^2t)}} - \int_0^a \frac{dt}{\sqrt{4t(1-t)(1-k^2t)}}.$$

This last question is solved with the help of Euler's theorem (the addition theorem for elliptic functions).

§35. Reduction of the general problem

In correspondence with the considerations in §34 we take the equation

$$\int_0^x \frac{dt}{\sqrt{4t(1-t)(1-k^2t)}} = M \int_0^y \frac{dt}{\sqrt{4t(1-t)(1-\lambda^2t)}}. \tag{1}$$

Setting

$$\int_0^x \frac{dt}{\sqrt{4t(1-t)(1-k^2t)}} = u,$$

we replace (1) by the parametric equations

$$\left.\begin{aligned} x &= \operatorname{sn}^2(u; k) \equiv \varphi(u), \\ y &= \operatorname{sn}^2(u/M; \lambda) \equiv \psi(u). \end{aligned}\right\} \tag{2}$$

Our problem now consists in the determination of conditions under which the two elliptic functions $\varphi(u)$ and $\psi(u)$ are connected by an algebraic relation.

Through passage to the parametric equations (2), the conditions we want might be dependences between the parameters k, λ, and M. We can now look for these conditions either in the form of dependences between the parameters k, λ, and M, or in the form of dependences between the periods 2ω and $2\omega'$ of $\varphi(u)$ and the periods $2\widetilde{\omega}$ and $2\widetilde{\omega}'$ of $\psi(u)$.

We shall see that, in contrast to the complicated dependences between k, λ, and M, the dependences between the periods have a simple and easily visible form, thanks to which it is possible to split the problem at the beginning into certain special problems, and then to solve these special problems by general methods in the theory of elliptic functions.

Accordingly, suppose that there is an algebraic dependence $F(\varphi(u), \psi(u)) = 0$ between the functions $x = \varphi(u)$ and $y = \psi(u)$. Replacing u by $u + 2m\omega'$ in this identity, where $m = \pm 1, \pm 2, \ldots$, we get an infinite set of identities

$$F(\varphi(u), \psi(u + 2m\omega')) = 0.$$

If $F(x, y) = 0$ is an equation of the nth degree in y, then for given x at most n distinct values are obtained for y. Consequently, among the values of $\psi(u)$ at the points

$$v, v \pm 2\omega', v \pm 4\omega', \ldots \tag{3}$$

there are at most n distinct numbers. This is possible only in the case when among the numbers (3) there are some congruences modulo the periods $2\widetilde{\omega}$ and $2\widetilde{\omega}'$. Assuming that

$$v + 2b\omega' = v + 2a\omega' + 2\alpha\widetilde{\omega}' + 2\beta\widetilde{\omega},$$

where a, b, α, and β are integers, we get

$$r\omega' = \alpha\widetilde{\omega}' + \beta\widetilde{\omega}$$

and, similarly,

$$s\omega = \gamma'\widetilde{\omega}' + \delta\widetilde{\omega}.$$

Thus, we have arrived at the following result: *if between the functions* (2) *there is an algebraic relation, then the periods of these functions are connected by the relations*

$$r\omega' = \alpha\widetilde{\omega}' + \beta\widetilde{\omega}, \qquad s\omega = \gamma\widetilde{\omega}' + \delta\widetilde{\omega}, \tag{4}$$

where $r, s, \alpha, \beta, \gamma$, and δ are integers.

The converse is also true: *if between the periods of the functions* (2) *there are dependences* (4) *with integer coefficients, then the functions* (2) *are connected by an algebraic relation.*

For a proof let $\Omega' = r\omega'$ and $\Omega = s\omega$, so that

$$\begin{cases} \Omega' = \alpha\widetilde{\omega}' + \beta\widetilde{\omega}, \\ \Omega = \gamma\widetilde{\omega}' + \delta\widetilde{\omega}. \end{cases}$$

We consider the function $\Phi(u) = \wp(u|\Omega, \Omega')$. Since each of the functions $\varphi(u)$ and $\psi(u)$ is even and has periods 2Ω and $2\Omega'$, each of these functions can be expressed rationally in terms of $\Phi(u)$:

$$\varphi(u) = R_1(\Phi(u)), \qquad \psi(u) = R_2(\Phi(u)).$$

Eliminating $\Phi(u)$ from these relations, we get an algebraic dependence between $\varphi(u)$ and $\psi(u)$. Thus, our assertion is proved.

In passing we have proved an important property of the algebraic equations

$$F(x,y) = 0 \tag{5}$$

to which (1) reduces, i.e., of the equations of the theory of transformations of elliptic functions: these equations admit a parametric representation

$$x = R_1(z), \qquad y = R_2(z), \tag{6}$$

where R_1 and R_2 are rational functions.

Algebraic curves (5) whose points have coordinates admitting a representation (6) bear the name of *unicursal* curves. They have many remarkable properties.

Our considerations have the important consequence that we can confine ourselves to the study of certain rational transformations alone. These transformations correspond to transformations of periods that can be expressed by the formulas

$$\omega' = \alpha\tilde{\omega}' + \beta\tilde{\omega}, \qquad \omega = \gamma\tilde{\omega}' + \delta\tilde{\omega},$$

where α, β, γ, and δ are integers, and the determinant $\left|\begin{smallmatrix} \alpha & \beta \\ \gamma & \delta \end{smallmatrix}\right| = n$ will always be assumed to be positive.

The area of a parallelogram of the periods 2ω and $2\omega'$ is n times larger than the area of a parallelogram of the periods $2\tilde{\omega}$ and $2\tilde{\omega}'$. We call the transformation under consideration an *nth-degree transformation*. We were concerned with a first-degree transformation ($n = 1$) in §§8 and 9 in the study of the modular group and modular functions as well as in §22, where the transformation of theta functions was studied. In §9 we established, in particular, that an arbitrary first-degree transformation is the result of successive application of two basic (or principal) first-degree transformations:

$$\omega' = \tilde{\omega}' + \tilde{\omega}, \qquad \omega = \tilde{\omega}; \tag{S}$$

$$\omega' = -\tilde{\omega}, \qquad \omega = \tilde{\omega}'. \tag{T}$$

It turns out that *an arbitrary nth-degree transformation can be obtained by repeatedly applying first-degree transformations and the following two transformations (so-called mth-degree principal transformations):*

$$\begin{cases} \omega' = \widetilde{\omega}' \\ \omega = m\widetilde{\omega} \end{cases} \quad or \quad \begin{cases} \widetilde{\omega}' = \omega', \\ \widetilde{\omega} = \omega/m \end{cases} \tag{I}$$

and

$$\begin{cases} \omega' = m\widetilde{\omega}', \\ \omega = \widetilde{\omega} \end{cases} \quad or \quad \begin{cases} \widetilde{\omega}' = \omega'/m, \\ \widetilde{\omega} = \omega, \end{cases} \tag{II}$$

where m is a positive integer. The first of these transformations consists in division of the first period (2ω) by the integer m, and the second consists in division of the second period $(2\omega')$.

Let us prove this assertion. Given the transformation

$$\omega' = \alpha\widetilde{\omega}' + \beta\widetilde{\omega}, \qquad \omega = \gamma\widetilde{\omega}' + \delta\widetilde{\omega},$$

we denote by r the greatest common divisor of the numbers α and β so that $\alpha = ra$ and $\beta = rb$, where a and b are relatively prime numbers. There exist integers c and d such that $ad - bc = 1$. Let

$$\omega_1' = a\widetilde{\omega}' + b\widetilde{\omega}, \qquad \omega_1 = c\widetilde{\omega}' + d\widetilde{\omega}.$$

Then

$$\omega' = r\omega_1', \qquad \omega = q\omega_1' + s\omega_1. \tag{7}$$

The transition from the pair $(\widetilde{\omega}, \widetilde{\omega}')$ to the pair (ω_1, ω_1') is realized by a first-degree transformation. Therefore, we can confine ourselves to the transformation (7). Taking $r > 0$ for definiteness, we let

$$\begin{cases} \omega' = r\omega_3', \\ \omega = \omega_3, \end{cases} \quad \begin{cases} \omega_2' = \omega_1', \\ \omega_2 = s\omega_1. \end{cases}$$

Here we have transformations of division of the second period in the pair $(2\omega, 2\omega')$ and of the first period in the pair $(2\omega_2, 2\omega_2')$. With the help of these transformations (7) takes the form

$$\omega_3' = \omega_2', \qquad \omega_3 = q\omega_2' + \omega_3,$$

which is a first-degree transformation.

We see that the transition from the periods $(2\omega, 2\omega')$ to the periods $(2\widetilde{\omega}, 2\widetilde{\omega}')$ is split into the following transformations: division of the second period, then a first-degree transformation, then division of the first period, and, finally, again a first-degree transformation.

Thus, our assertion is proved.

It is not hard to see that $n = r \cdot s$. We see also that in considering (I) and (II) we can assume that m is a prime number, and that in any case we can consider separately the case of odd m and the case $m = 2$.

Strictly speaking, it is not necessary to consider both the transformations (I) and (II), since, for example, (II) can be reduced to (I) and first-degree transformations.

We illustrate this remark in §39. However, it is more advantageous to consider (II) directly without reducing it to other transformations.

§36. The first principal first-degree transformation

Let us consider the functions

$$x = \mathrm{sn}^2(u; k), \qquad y = \mathrm{sn}^2(u/M; \lambda).$$

The numbers $2K$ and $2iK'$ are the periods of $\mathrm{sn}^2(u; k)$. Similarly, denote by $2L$ and $2iL'$ the periods of $\mathrm{sn}^2(v; \lambda)$. Thus, y has periods $2ML$ and $2iML'$. For the first transformation of first degree,

$$iML' = iK' + K, \qquad ML = K.$$

Consider the ratio y/x. This is an even elliptic function with periods $2K$ and $2iK'$. Consequently, it can be expressed rationally in terms of $\mathrm{sn}^2(u; k)$. In the parallelogram of the periods $2K$ and $2iK'$ the function y/x has a two-fold pole at the point $u = iML' = iK' + K$ since the numerator y has a second-order pole there and the denominator is nonzero and finite there. Further, y/x has a second-order zero at the point $u = iK'$, because at this point the numerator is finite and nonzero, while the denominator has a second-order pole. By the foregoing,

$$\frac{y}{x} = \frac{C}{\mathrm{sn}^2(u; k) - \mathrm{sn}^2(iK' + K; k)},$$

or

$$\frac{y}{x} = \frac{A}{1 - k^2 x},$$

where A, like C, is a constant. To determine this constant, let $u = K$. This gives us that $1 = A/(1 - k^2)$. Accordingly, $A = k'^2$, and hence

$$\frac{\mathrm{sn}^2(u/M; \lambda)}{\mathrm{sn}^2(u; k)} = \frac{k'^2}{1 - k^2 \mathrm{sn}^2(u; k)}, \tag{1}$$

i.e.,

$$\mathrm{sn}\left(\frac{u}{M}; \lambda\right) = \frac{k' \mathrm{sn}(u; k)}{\mathrm{dn}(u; k)}.$$

It remains to express M and λ in terms of k. Dividing this equality by u and letting u go to zero, we get that $1/M = k'$. Further, setting $u = 2K + iK'$ in (1), we get that $1/\lambda^2 = -k'^2/k^2$. Consequently, $\lambda = ik/k'$.

Thus,

$$\mathrm{sn}\left(k'u; \frac{ik}{k'}\right) = \frac{k' \mathrm{sn}(u; k)}{\mathrm{dn}(u; k)}.$$

It can be proved similarly (or with the help of (1)) that

$$\mathrm{cn}\left(k'u;\frac{ik}{k'}\right) = \frac{\mathrm{cn}(u;k)}{\mathrm{dn}(u;k)},$$

$$\mathrm{dn}\left(k'u;\frac{ik}{k'}\right) = \frac{1}{\mathrm{dn}(u;k)}.$$

Let k vary in the interval from 0 to 1. As we have already mentioned, this is the most important case for applications. The quantity

$$k/\sqrt{1-k^2} = k/k',$$

then varies monotonically from 0 to ∞. Thus, the formulas obtained in this section carry Jacobi functions with purely imaginary modulus into Jacobi functions with modulus in $[0,1]$.

We remark that the transformation formulas found in this section could have been obtained from the transformation formulas for theta functions derived in §22. The same remark relates to the formulas obtained in the next section.

§37. The second principal first-degree transformation

This transformation corresponds to the following relations:

$$ML = iK', \qquad iML' = -K.$$

Further, as above, $2K$ and $2iK'$ are the periods of the function

$$x = \mathrm{sn}^2(u;k),$$

and $2ML$ and $2iML'$ are the periods of the function

$$y = \mathrm{sn}^2(u/M;\lambda).$$

We again consider the ratio y/x. This is a rational function of $\mathrm{sn}^2(u;k)$ that has in the parallelogram of the periods $2K$ and $2iK'$ a two-fold zero at the point $u = iK'$ and a two-fold pole at the point $u = K$. Therefore,

$$\frac{y}{x} = \frac{A}{\mathrm{sn}^2(u;k) - 1}.$$

Letting u go to zero, we get $1/M^2 = -A$. We have that

$$y = \frac{A\,\mathrm{sn}^2(u;k)}{\mathrm{sn}^2(u;k) - 1};$$

therefore, setting $u = iK'$, we find that $1 = A$.

Replacing u by $-K + iK'$, we get

$$\mathrm{sn}^2(L + iL';\lambda) = \frac{\mathrm{sn}^2(K + iK';k)}{\mathrm{sn}^2(K + iK';k) - 1},$$

or

$$\frac{1}{\lambda^2} = \frac{1/k^2}{1/k^2 - 1}.$$

This implies that $\lambda = k'$, and since $M = \pm i$, our result is that

$$\mathrm{sn}(iu; k') = i\frac{\mathrm{sn}(u; k)}{\mathrm{cn}(u; k)}. \tag{1}$$

It can be proved similarly (or on the basis of (1)) that

$$\mathrm{cn}(iu; k') = \frac{1}{\mathrm{cn}(u; k)}, \qquad \mathrm{dn}(iu; k') = \frac{\mathrm{dn}(u; k)}{\mathrm{cn}(u; k)}.$$

The value of these formulas consists in the fact that they carry Jacobi functions of a purely imaginary argument into Jacobi functions of a real argument.

§38. Landen's transformation

The first principal second-degree transformation was discovered as far back as 1775 by Landen.[18] The scheme corresponding to this transformation is

$$x = \mathrm{sn}^2(u; k); \qquad 2K, \qquad 2iK',$$
$$y = \mathrm{sn}^2(u/M; \lambda); \quad 2ML = K, \quad 2iML' = 2iK',$$

from which it is clear that this involves division of the first period of x by two.

The ratio y/x is a rational function of $\mathrm{sn}^2(u; k)$. This function has a two-fold zero at the point $u = K$ and a two-fold pole at the point $u = K + iK'$ in the parallelogram of the periods $2K$ and $2iK'$. Therefore,

$$\frac{y}{x} = C\frac{\mathrm{sn}^2(K; k) - \mathrm{sn}^2(u; k)}{\mathrm{sn}^2(K + iK'; k) - \mathrm{sn}^2(u; k)}$$

or

$$\frac{y}{x} = \frac{A(1 - x)}{1 - k^2 x}.$$

Setting $u = 0$, iK', and $K/2$ successively, we get the following equalities:

$$\frac{1}{M^2} = A, \qquad \frac{k^2 M^2}{\lambda^2} = \frac{A}{k^2},$$
$$1 = \frac{A\,\mathrm{sn}^2(K/2; k)\mathrm{cn}^2(K/2; k)}{\mathrm{dn}^2(K/2; k)} \tag{1}$$

---

[18]Landen, of course, was studying not elliptic functions but elliptic integrals.

The last of these can be reduced to the form $A = (1 + k')^2$ if we use (see the exercise at the end of §27) the formula

$$\mathrm{sn}(K/2; k) = 1/\sqrt{1 + k'}.$$

The first two equations in (1) now give us that

$$M = \frac{1}{1 + k'}, \qquad \lambda = \frac{k^2}{(1 + k')^2} = \frac{1 - k'}{1 + k'}.$$

Our result is that

$$\mathrm{sn}\left[u(1 + k'); \frac{1 - k'}{1 + k'}\right] = \frac{(1 + k')\mathrm{sn}(u; k)\mathrm{cn}(u; k)}{\mathrm{dn}(u; k)}.$$

The rest of the formulas have the form

$$\mathrm{cn}\left[u(1 + k'); \frac{1 - k'}{1 + k'}\right] = \frac{1 - (1 + k')\mathrm{sn}^2(u; k)}{\mathrm{dn}(u; k)},$$

$$\mathrm{dn}\left[u(1 - k'); \frac{1 - k'}{1 + k'}\right] = \frac{1 - (1 - k')\mathrm{sn}^2(u; k)}{\mathrm{dn}(u; k)},$$

Their verification is left to the reader.

§39. Gauss's transformation

This transformation consists in division of the second period by two. Therefore, it can be obtained by combining Landen's transformation (consisting of division of the first period by two) with first-degree transformations.

As above, the original modulus is k, and we come to the modulus λ after the transformation. We take the formulas of the first-degree transformations that interchange the old and new periods. They have the form

$$\mathrm{sn}(iu; k') = i\frac{\mathrm{sn}(u; k)}{\mathrm{cn}(u; k)},$$

$$\mathrm{cn}(iu; k') = \frac{1}{\mathrm{cn}(u; k)}, \tag{1}$$

$$\mathrm{dn}(iu; k') = \frac{\mathrm{dn}(u; k)}{\mathrm{cn}(u; k)};$$

and

$$\mathrm{sn}\left(\frac{iu}{M}; \lambda'\right) = i\frac{\mathrm{sn}(u/M; \lambda)}{\mathrm{cn}(u/M; \lambda)},$$

$$\mathrm{cn}\left(\frac{iu}{M}; \lambda'\right) = \frac{1}{\mathrm{cn}(u/M; \lambda)}, \tag{2}$$

$$\mathrm{dn}\left(\frac{iu}{M}; \lambda'\right) = \frac{\mathrm{dn}(u/M; \lambda)}{\mathrm{cn}(u/M; \lambda)}.$$

Here λ and M are arbitrary so far. We must now require that the left-hand sides of (1) and (2) be connected by Landen's transformation. This determines λ and M.

Recalling Landen's transformation and taking into account that k' and λ' now play the role of the previous k and λ, we find at once that

$$\lambda' = \frac{1-k}{1+k}, \qquad M = \frac{1}{1+k}$$

and

$$\lambda = \sqrt{1 - \lambda'^2} = \frac{2\sqrt{k}}{1+k}.$$

Thus, both parameters λ and M are determined. The transformation formulas have the form

$$\operatorname{sn}\left(\frac{iu}{M}; \lambda'\right) = \frac{1}{M} \frac{\operatorname{sn}(iu; k')\operatorname{cn}(iu; k)}{\operatorname{dn}(iu; k')},$$

$$\operatorname{cn}\left(\frac{iu}{M}; \lambda'\right) = \frac{1 - (1+k)\operatorname{sn}^2(iu; k')}{\operatorname{dn}(iu; k)},$$

$$\operatorname{dn}\left(\frac{iu}{M}; \lambda'\right) = \frac{1 - (1-k)\operatorname{sn}^2(iu; k')}{\operatorname{dn}(iu'k')}.$$

It now remains to use (1) and (2). This gives us that

$$\frac{\operatorname{su}(u/M; \lambda)}{\operatorname{cn}(u/M; \lambda)} = \frac{1}{M} \frac{\operatorname{sn}(u; k)}{\operatorname{cn}(u; k)\operatorname{dn}(u; k)},$$

$$\frac{1}{\operatorname{cn}(u/M; \lambda)} = \frac{\operatorname{cn}^2(u; k) + (1+k)\operatorname{sn}^2(u; k)}{\operatorname{cn}(u; k)\operatorname{dn}(u; k)},$$

$$\frac{\operatorname{dn}(u/M; \lambda)}{\operatorname{cn}(u/M; \lambda)} = \frac{\operatorname{cn}^2(u; k) + (1-k)\operatorname{sn}^2(u; k)}{\operatorname{cn}(u; k)\operatorname{dn}(u; k)}$$

and the final formulas have the form

$$\operatorname{sn}\left(\frac{u}{M}; \lambda\right) = \frac{1}{M} \frac{\operatorname{sn}(u; k)}{1 + k\operatorname{sn}^2(u; k)},$$

$$\operatorname{cn}\left(\frac{u}{M}; \lambda\right) = \frac{\operatorname{cn}(u; k)\operatorname{dn}(u; k)}{1 + k\operatorname{sn}^2(u; k)},$$

$$\operatorname{dn}\left(\frac{u}{M}; \lambda\right) = \frac{1 - k\operatorname{sn}^2(u; k)}{1 + k\operatorname{sn}^2(u; k)}.$$

We have reduced Gauss's transformation to Landen's transformation intentionally in order to illustrate the remark made at the end of §35. A direct treatment would have led to the final formulas more quickly.

§40. Principal transformations of nth degree

These transformations consist in division of one of the periods by n. The method for obtaining the formulas corresponding to this transformation was explained well enough in the preceding sections. Therefore, we confine ourselves here to a detailed analysis of only one case. The reader can find a complete summary of the formulas relating to nth-degree transformations in Tables XXII and XXIII.

Let us consider division of the second period by n, where n can be either even or odd. We have the following scheme:

$$L = K/M, \qquad L' = K'/nM.$$

$$x = \operatorname{sn}(u; k), \qquad y = \operatorname{sn}(u/M; \lambda).$$

The ratio y/x is an even function of u and, as is easily seen, has periods $2K$ and $2iK'$, since

$$\operatorname{sn}(u + 2mK + 2m'iK'; k) = (-1)^m \operatorname{sn}(u; k),$$

$$\operatorname{sn}\left(\frac{u + 2mK + 2m'iK'}{M}; \lambda\right) = (-1)^m \operatorname{sn}\left(\frac{u}{M}; \lambda\right).$$

Consequently, y/x is a rational function of $\operatorname{sn}^2(u; k)$. In the fundamental parallelogram of the periods $2K$ and $2iK'$ the function under consideration has simple zeros at the points

$$u = \frac{i\mu K'}{n} \qquad \left(\mu = \pm 2, \pm 4, \ldots, \pm 2\left[\frac{n}{2}\right]\right)$$

and simple poles at the points

$$u = \frac{i\nu K'}{n} \qquad \left\{\nu = \pm 1, \pm 3, \ldots, \pm\left(2\left[\frac{n}{2}\right] - 1\right)\right\}.$$

From this,

$$\frac{\operatorname{sn}(u/M; \lambda)}{\operatorname{sn}(u; k)}$$

$$= \frac{1}{M} \prod_{r=1}^{[n/2]} \frac{1 - \operatorname{sn}^2(u; k)/\operatorname{sn}^2(2irK'/n; k)}{1 - \operatorname{sn}^2(u; k)\operatorname{sn}^2((2r-1)iK'/n; k)}. \tag{1}$$

To determine M, let $u = K$. This gives us

$$M = \prod_{r=1}^{[n/2]} \frac{\operatorname{cn}^2(2irK'/n; k)\operatorname{sn}^2((2r-1)iK'/n; k)}{\operatorname{sn}^2(2irK'/n; k)\operatorname{cn}^2((2r-1)iK'/n; k)}.$$

Observing that

$$\operatorname{sn}(iu;k)/\operatorname{cn}(iu;k) = i\operatorname{sn}(u;k'), \tag{2}$$

we can rewrite the expression just obtained as follows:

$$M = \prod_{r=1}^{[n/2]} \frac{\operatorname{sn}^2((2r-1)K'/n;k')}{\operatorname{sn}^2(2rK'/n;k')}. \tag{3}$$

To determine λ we set $u = K + iK'/n$ in (1). Since

$$\operatorname{sn}\left(K + \frac{iK'}{n};k\right) = \frac{1}{\operatorname{dn}(K'/n;k')},$$

$$\operatorname{sn}(L + iL';\lambda) = 1/\lambda,$$

we again use (2) to get that

$$\frac{\operatorname{dn}(K'/n;k')}{\lambda}$$

$$= \frac{1}{M} \prod_{r=1}^{[n/2]} \frac{1 + \operatorname{cn}^2(2rK'/n;k')/\operatorname{dn}^2(K'/n;k')\operatorname{sn}^2(2rK'/n;k')}{1 + \operatorname{cn}^2((2r-1)K'/n;k')/\operatorname{dn}^2(K'/n;k')\operatorname{sn}^2((2r-1)K'/n;k')}.$$

Observe now that

$$\operatorname{cn}^2\alpha + \operatorname{dn}^2\beta\operatorname{sn}^2\alpha = 1 - k^2\operatorname{sn}^2\alpha\operatorname{sn}^2\beta$$

$$= \frac{\Theta^2(0)\Theta(\alpha + \beta)\Theta(\alpha - \beta)}{\Theta^2(\alpha)\Theta^2(\beta)}.$$

In view of this relation and (3),

$$\lambda = \operatorname{dn}\left(\frac{K'}{n};k'\right) \prod_{r=1}^{[n/2]} \frac{\Theta^3(2rK'/n|\tau')\Theta((2r-2)K'/n|\tau')}{\Theta^3((2r-1)K'/n|\tau')\Theta((2r+1)K'/n|\tau')}.$$

We now take into account that

$$\operatorname{dn}\left(\frac{K'}{n};k'\right) = \frac{\Theta((n \pm 1)K'/n|\tau')\Theta(0|\tau')}{\Theta(K'/n|\tau')\Theta(K'|\tau')}.$$

This enables us to give the modulus λ the following form:

$$\lambda = \left\{\prod_{r=1}^{n} \frac{\Theta(2rK'/n|\tau')}{\Theta((2r-1)K'/n|\tau')}\right\}^2$$

either for an even or for an odd n.

CHAPTER 7

Additional Facts About Elliptic Integrals

§41. Elliptic curves of general form

We consider the equation

$$w^2 = f(z), \tag{1}$$

where $f(z) = a_0 z^4 + 4a_1 z^3 + 6a_2 z^2 + 4a_3 z + a_4$ is an arbitrary polynomial of fourth degree without multiple roots. The set of all pairs (z, w) satisfying (1) is called an *elliptic curve*, and each pair (z, w) is a point of this curve.[19]

Associated with an elliptic curve are elliptic integrals, in the first place the integral of the first kind

$$u = \int_c^z \frac{dx}{\sqrt{f(x)}}, \tag{2}$$

instead of which we can consider the differential equation

$$dz/du = \sqrt{f(z)}. \tag{3}$$

In the special case when (1) has the canonical Weierstrass form, i.e.,

$$f(z) = 4z^3 - g_2 z - g_3,$$

we have already obtained the following results:

a) The curve (1) admits a so-called uniformization by means of elliptic functions, namely, if $z = \wp(u)$ and $w = \wp'(u)$, then the pair (z, w) runs through the whole curve when the parameter u runs through a period parallelogram.

b) In particular, the roots of the equation $f(z) = 0$ can be represented by means of elliptic functions, namely, $e_k = \wp(\omega_k)$, $k = 1, 2, 3$.

c) The problem of inversion of the integral (2) can be solved in elliptic functions; more precisely, the upper limit z in the formula (2) is equal to

---

[19] An elliptic curve is a special case of a general algebraic curve (see §54).

$\wp(u)$ if $c = \infty$, and is equal to $\wp(u + \omega_k)$ if $c = e_k$ $(k = 1, 2, 3)$, or, in other words, the functions $z = \wp(u)$ and $z = \wp(u + \omega_k)$ $(k = 1, 2, 3)$ represent solutions of (3) under the initial conditions $z(0) = \infty$ and $z(0) = e_k$ $(k = 1, 2, 3)$, respectively.

With the help of a suitable transformation

$$z = \frac{\alpha z_1 + \beta}{\gamma z_1 + \delta}, \quad w = (\gamma z_1 + \delta)^2 w_1 \quad (\alpha\delta - \beta\gamma = 1)$$

the curve (1) can be transformed into the canonical curve

$$w_1^2 = 4z_1^3 - g_2 z_1 - g_3;$$

therefore, all three of the facts above hold in a certain modified formulation for an arbitrary elliptic curve (1).

However, the necessary transformation requires knowledge of some root of the polynomial $f(z)$, and hence cannot lead to formulas useful for investigation if the coefficients of $f(z)$ do not have definite numerical values but are parameters.

In this section we show that these difficulties can be gotten around.

Since the case when $a_0 = 0$ is not interesting, we assume (clearly without loss of generality) that $a_0 = 1$. Further, we can assume that $a_1 = 0$, because this can be attained by using the transformation $z = z_1 - a_1$, which replaces the old coefficients by certain integral rational functions of them.

Accordingly, we assume that the given equation has the form

$$w^2 = z^4 - 6Az^2 + 4Bz + C, \tag{I}$$

where A, B, and C are assumed to be arbitrary coefficients.

We write the invariants

$$g_2 = C + 3A^2, \qquad g_3 = -AC + A^3 - B^2.$$

It follows from these formulas that

$$B^2 = 4A^3 - g_2 A - g_3, \tag{4}$$

$$C = g_2 - 3A^2. \tag{5}$$

Relation (4) shows that (A, B) is a point of an elliptic curve in the Weierstrass canonical form. Consequently, there is a v such that

$$A = \wp(v), \qquad B = \wp'(v). \tag{II}$$

Therefore, (5) gives us that

$$C = g_2 - 3[\wp(v)]^2. \tag{III}$$

We take as parameters the quantities g_2, g_3, and v instead of the original quantities A, B, and C. Equation (I) can be rewritten in the form

$$w^2 = [z^2 - 3\wp(v)]^2 + 4\wp'(v)z + g_2 - 12[\wp(v)]^2. \tag{I'}$$

We now recall the identity (9) in §15:

$$[\wp(u - v/2) - \wp(u + v/2)]^2$$
$$= \left\{ \frac{1}{4}\left[\frac{\wp'(u - v/2) - \wp'(v)}{\wp(u - v/2) - \wp(v)} \right]^2 - 3\wp(v) \right\}^2$$
$$+ 2\wp'(v)\frac{\wp'(u - v/2) - \wp'(v)}{\wp(u - v/2) - \wp(v)} + g_2 - 12[\wp(v)]^2.$$

Comparison with (I') shows that this equation is satisfied identically if we set

$$z = \frac{1}{2}\frac{\wp'(u - v/2) - \wp'(v)}{\wp(u - v/2) - \wp(v)}, \tag{IV}$$
$$w = \wp(u - v/2) - \wp(u + v/2). \tag{V}$$

It is not hard to see that the required construction is now complete.

We formulate the result.

Given an elliptic curve (I) with the invariants g_2 and g_3, we represent its coefficients according to (II) and (III). In this case a uniformization of the curve (I) is given by (IV) and (V); this relates to a).

To get the roots of the polynomial on the right-hand side of (I) we should set $w = 0$ in (V) and find the values of u. Substituting them in (IV), we get the following expressions for the roots:

$$z_0 = -\frac{1}{2}\frac{\wp'(v/2) + \wp'(v)}{\wp(v/2) - \wp(v)} = \varphi(v), \tag{VI}$$
$$z_k = \varphi(v + 2\omega_k) \qquad (k = 1, 2, 3).$$

This relates to b).

Finally, the problem of inversion is solved by each of the formulas

$$z = \frac{1}{2}\frac{\wp'(u - v/2) - \wp'(v)}{\wp(u - v/2) - \wp(v)} = \varphi(u, v), \tag{VII}$$
$$z = \varphi(u, v + 2\omega_k) \qquad (k = 1, 2, 3).$$

Indeed, we prove that $z = \varphi(u, v)$ satisfies (3) as a function of u (with value z_0 at $u = 0$). But this follows from the addition theorem for the function $\wp$ (see Table VI):

$$\frac{d}{du}\varphi(u, v) = \frac{1}{2}\frac{\partial}{\partial u}\frac{\wp'(u - v/2) - \wp'(v)}{\wp(u - v/2) - \wp(v)}$$
$$= \wp(u - v/2) - \wp(u + v/2) = w.$$

Similarly, $z = \varphi(u, v + 2\omega_k)$ is the solution of (3) with value z_k at $u = 0$.

Exercise. Prove that (IV) can be replaced by

$$z = \frac{1}{2}\frac{\wp'(u - v/2) + \wp'(u + v/2)}{\wp(u - v/2) - \wp(u + v/2)}, \tag{IV'}$$

and (VI) by

$$z_0 = -\frac{1}{2}\frac{\wp''(v/2)}{\wp'(v/2)} = \psi(v),$$

$$z_k = \psi(v + 2\omega_k) \qquad (k = 1, 2, 3). \tag{VI'}$$

§42. The function $\wp(u)$ with real invariants

If the invariants g_2 and g_3 are real, then either all three roots e_1, e_2, and e_3 of the polynomial $4z^3 - g_2 z - g_3$ are real, or one of them is real (say e_2) and the other two are complex conjugates.

In the first case, which holds if the discriminant is positive ($g_2^3 - 27g_3^2 > 0$), we remember the roots in such a way that $e_1 > e_2 > e_3$. It should be kept in mind that the roots are assumed to be distinct; this excludes, by the way, the vanishing of the discriminant.

The second case holds if the discriminant is negative.

These facts are an immediate consequence of the formula

$$g_2^3 - 27g_3^2 = 16(e_1 - e_2)^2(e_2 - e_3)^2(e_3 - e_1)^2.$$

A function $\wp(u)$ with real invariants has only real values on the real axis (for either sign of the discriminant). Indeed, in the case of real invariants all the coefficients in the expansion

$$\wp(u) = \frac{1}{u^2} + \frac{g_2}{20}u^2 + \frac{g_3}{28}u^4 + \cdots$$

are real.

Similarly, the values taken by $\wp(u)$ on the imaginary axis are real; here the relation

$$\wp(iu; g_2, g_3) = -\wp(u; g_2, -g_3) \tag{1}$$

is useful for the computation.

We now occupy ourselves with the derivation of expressions for the periods $2\omega_1$ and $2\omega_3$ of $\wp(u)$ in terms of invariants.

1°. Let us begin with the case when the discriminant is positive. The formula

$$u = \int_\wp^\infty \frac{dx}{\sqrt{4x^3 - g_2 x - g_3}},$$

where for $x > e_1$ the square root is understood to be its arithmetic value, implies that, as $\wp$ decreases from ∞ to e_1, the quantity u increases monotonically from 0 to ω_1. Consequently,

$$\omega_1 = \int_{e_1}^{\infty} \frac{dx}{\sqrt{4x^3 - g_2 x - g_3}}. \tag{2}$$

We now take the function $\wp(u; g_2, -g_3)$ and denote its periods by $2\widetilde{\omega}_1$ and $2\widetilde{\omega}_3$. The roots of the polynomial $4x^3 - g_2 x + g_3$, in decreasing order, are $\widetilde{e}_1 = -e_3, \widetilde{e}_2 = -e_2$, and $\widetilde{e}_3 = -e_1$. Further, relation (1) shows that

$$\widetilde{\omega}_1 = \omega_3/i, \qquad \widetilde{\omega}_3 = -\omega_1/i. \tag{3}$$

By analogy with (2) we can write

$$\widetilde{\omega}_1 = \int_{\widetilde{e}_1}^{\infty} \frac{dx}{\sqrt{4x^3 - g_2 x + g_3}}.$$

Consequently,

$$\omega_3 = i \int_{-e_3}^{\infty} \frac{dx}{\sqrt{4x^3 - g_2 x + g_3}}. \tag{4}$$

Thus, the periods $2\omega_1$ and $2\omega_3$ are expressed in terms of the invariants g_2 and g_3 as integrals. We see that $2\omega_1$ is a real number, while $2\omega_3$ is a purely imaginary number. This means that in this case of a positive discriminant the period parallelogram is a rectangle.

Using the connection between Weierstrass functions and Jacobi functions, we can write

$$\omega = K/\sqrt{e_1 - e_3}, \qquad \omega' = iK'/\sqrt{e_1 - e_3},$$

where

$$\sqrt{e_1 - e_3} > 0, \qquad k^2 = (e_2 - e_3)/(e_1 - e_3).$$

$2°$. We proceed now to the case of a negative discriminant. Here we get by analogous considerations the formula

$$\omega_2 = \int_{e_2}^{\infty} \frac{dx}{\sqrt{4x^3 - g_2 x - g_3}}. \tag{5}$$

Let us again take the function $\wp(u; g_2, -g_3)$. Keeping the previous notation, we get the analogous formula

$$\widetilde{\omega}_2 = \int_{-e_2}^{\infty} \frac{dx}{\sqrt{4x^3 - g_2 x + g_3}}. \tag{6}$$

Consider now (3) and the fact that

$$\omega_2 = -\omega_1 - \omega_3, \qquad \widetilde{\omega}_2 = -\widetilde{\omega}_1 - \widetilde{\omega}_3.$$

We find that

$$\omega_1 + \omega_3 = -\omega_2, \qquad \omega_1 - \omega_3 = i\widetilde{\omega}_2.$$

Thus, adding and subtracting the quantities (5) and (6), we get the periods $2\omega_1$ and $2\omega_3$. We see that in the case of a negative discriminant the periods $2\omega_1$ and $2\omega_3$ are conjugate numbers, which implies that the fundamental parallelogram is a rhombus.

It is also possible to express the periods in terms of normal elliptic integrals in Legendre form in the case of a negative discriminant. With this goal we should set

$$e_2 - e_3 = \rho(\cos\psi + i\sin\psi),$$

where $\rho > 0$ and $0 < \psi < \pi$. It can be shown[20] that in this notation

$$\omega_2 = K/\sqrt{\rho}, \qquad \widetilde{\omega}_2 = K'/\sqrt{\rho}$$

where $\sqrt{\rho} > 0$ and $k^2 = \sin^2(\psi/2)$.

§43. Reduction of elliptic integrals to Jacobi normal form in the real case

We consider the integral

$$\int R(z, w)\, dz, \tag{1}$$

where R denotes a rational function of its arguments, and

$$w^2 = a_0 z^4 + 4a_1 z^3 + 6a_2 z^2 + 4a_3 z + a_4 \equiv f(z).$$

It is assumed here that the coefficients a_i are real numbers with $a_0 \neq 0$, that the polynomial $f(z)$ (which does not have multiple roots) is positive in some intervals of the number line, and z varies in these intervals. Thus, w has only real values. If, therefore, the coefficients of $R(z, w)$ are real, the integral (1) is also real. In this case (which we call the real case) it is

---

[20]To do this make the change of variable $x - e_2 = \rho t^2$ in (5), obtaining

$$\sqrt{\rho}\,\omega_2 = \int_0^\infty \frac{dt}{\sqrt{(t^2 + 1)^2 - 4k^2 t^2}} = \int_0^1 + \int_1^\infty .$$

Replacing t by $1/t$ in the second integral on the right-hand side gives us

$$\sqrt{\rho}\,\omega_2 = 2\int_0^1 \frac{dt}{\sqrt{(t^2 + 1)^2 - 4k^2 t^2}}.$$

The change of variable $t = \tan(\varphi/2)$ here gives us

$$\sqrt{\rho}\,\omega_2 = \int_0^1 \frac{d\varphi}{\sqrt{1 - k^2 \sin^2\varphi}} = K.$$

desirable to reduce the integral to normal form by a real transformation. Such a transformation is possible and is the object of this section.

First of all, note that by a decomposition into linear factors with a subsequent grouping of them it is possible to bring the polynomial $f(z)$ to the form

$$f(z) = a_0(z^2 + 2\lambda z + \mu)(z^2 + 2\rho z + \sigma),$$

where all the coefficients a_0, λ, μ, ρ, and σ are again real. If not all four roots of $f(z)$ are real, then this decomposition is unique, since at least one of the quadratic trinomials must have conjugate roots. But if all the roots are real, then the required decomposition is not unique. We then choose the one for which the roots of the first trinomial are larger than the roots of the second.

We can now conceive of two cases, depending on whether $\lambda = \rho$ or $\lambda \neq \rho$.

If $\lambda = \rho$, then let $z + \lambda = t$, and $f(z)$ takes the form

$$f(z) = a_0(t^2 + \alpha)(t^2 + \beta),$$

where α and β are real numbers.

But if $\lambda \neq \rho$, then we introduce instead of z a variable t according to the formula $z = (pt + q)/(t + 1)$, where p and q are as yet undetermined numbers. The polynomial $f(z)$ takes the form

$$f(z) = \frac{a_0}{(t+1)^4}\{(pt + q)^2 + 2\lambda(pt + q)(t + 1) + \mu(t + 1)^2\}$$
$$\times \{(pt + q)^2 + 2\rho(pt + q)(t + 1) + \sigma(t + 1)^2\}.$$

We require that the quadratic trinomials in the curly brackets do not contain the first power of the variable t. This leads to the following two equations for determining p and q:

$$\left.\begin{aligned} pq + \lambda(p + q) + \mu = 0, \\ pq + \rho(p + q) + \sigma = 0. \end{aligned}\right\} \tag{2}$$

We prove that we get real values when we solve these equations. It follows from (2) that

$$p + q = -\frac{\mu - \sigma}{\lambda - \rho}, \qquad pq = -\frac{\lambda\sigma - \mu\rho}{\lambda - \rho}.$$

Therefore, p and q are the roots of the quadratic equation

$$X^2 + \frac{\mu - \sigma}{\lambda - \rho}X - \frac{\lambda\sigma - \mu\rho}{\lambda - \rho} = 0.$$

The discriminant of this equation is equal to

$$D = \left(\frac{\mu - \sigma}{\lambda - \rho}\right)^2 + 4\frac{\lambda\sigma - \mu\rho}{\lambda - \rho}$$
$$= \frac{(\mu - \sigma)^2 + 4(\lambda - \rho)(\lambda\sigma - \mu\rho)}{(\lambda - \rho)^2}.$$

Denote the roots of the polynomial $z^2 + 2\lambda z + \mu$ by a and b, and the roots of $z^2 + 2\rho z + \sigma$ by c and d. Then

$$\mu = ab, \qquad 2\lambda = -a - b,$$
$$\sigma = cd, \qquad 2\rho = -c - d,$$

and hence

$$D = \frac{(ab - cd)^2 + (c + d - a - b)\{(c + d)ab - (a + b)cd\}}{(\lambda - \rho)^2}$$
$$= \frac{(a - c)(a - d)(b - c)(b - d)}{(\lambda - \rho)^2}.$$

If all the roots of $f(z)$ are real, then this expression is positive, since the numbers a and b are greater than c and d by assumption. The quantity D is positive also when not all the roots of $f(z)$ are real. For example, if c and d are real but a and b are not, then

$$(a - c)(b - c) = |a - c|^2,$$
$$(a - d)(b - d) = |a - d|^2.$$

But if all the roots are not real, then

$$(a - c)(b - d) = |a - c|^2,$$
$$(a - d)(b - c) = |a - d|^2.$$

Thus, D is always positive. Hence, p and q are real, and we have proved that our integral (1) can be reduced by a real linear fractional transformation to the form

$$\int R(t, \sqrt{\pm(t^2 + \alpha)(t^2 + \beta)})\, dt,$$

where R denotes some new rational function. We made the assumption that $f(z)$ is positive on some intervals of the number line. Consequently, the expression $\pm(t^2 + \alpha)(t^2 + \beta)$ also has this property. Therefore, the radical

$$y = \sqrt{\pm(t^2 + \alpha)(t^2 + \beta)}$$

can be of only the following types:

$$y = \sqrt{(a^2 - t^2)(b^2 - t^2)}, \tag{I}$$

$$y = \sqrt{(a^2 - t^2)(t^2 - b^2)}, \tag{II}$$

$$y = \sqrt{(a^2 - t^2)(b^2 + t^2)}, \tag{III}$$

$$y = \sqrt{(t^2 - a^2)(b^2 + t^2)}, \tag{IV}$$

$$y = \sqrt{(t^2 + a^2)(t^2 + b^2)}. \tag{V}$$

Here a and b denote positive numbers. Reduction to normal form does not present any difficulty in each of these cases. The corresponding formulas are contained in Table XXV.

Here we consider type I for an example. Thus, let

$$y = \sqrt{(a^2 - t^2)(b^2 - t^2)},$$

where $a^2 > b^2$. Two cases are possible, depending on whether $t^2 < b^2$ or $t^2 > a^2$. The inequality $b^2 < t^2 < a^2$ is excluded, because nonreal values of y correspond to it.

If $t^2 < b^2$, we set $t = bx$ and $k^2 = b^2/a^2$. This gives us

$$y = ab\sqrt{(1 - x^2)(1 - k^2x^2)}.$$

If $t^2 > a^2$, we set $t = a/x$ and $k^2 = b^2/a^2$. This gives us

$$y = \frac{a^2}{x^2}\sqrt{(1 - x^2)(1 - k^2x^2)}.$$

In both cases x varies in $[-1, 1]$, and it is convenient to set

$$x = \operatorname{sn}(u; k).$$

Then for u we have the interval $[-K, K]$.

§44. Complete elliptic integrals as hypergeometric functions

We take complete elliptic integrals in the trigonometric form

$$K = \int_0^{\pi/2} \frac{d\varphi}{\sqrt{1 - k^2 \sin^2 \varphi}}, \tag{1}$$

$$E = \int_0^{\pi/2} \sqrt{1 - k^2 \sin^2 \varphi}\, d\varphi, \tag{2}$$

and assume that $|k^2| < 1$. Under this assumption we can expand the integrands in series of increasing powers of the variable k^2:

$$\frac{1}{\sqrt{1 - k^2 \sin^2 \varphi}} = \sum_{r=0}^{\infty} \frac{(2r)!}{2^{2r}(r!)^2} k^{2r} \sin^{2r} \varphi,$$

$$\sqrt{1 - k^2 \sin^2 \varphi} = 1 - \sum_{r=1}^{\infty} \frac{(2r - 2)!}{2^{2r-1}(r - 1)!r!} k^{2r} \sin^{2r} \varphi.$$

Integrating these expressions and taking into account that

$$\int_0^{\pi/2} \sin^{2r} \varphi \, d\varphi = \frac{1}{2} \frac{\Gamma(r + 1/2)\Gamma(1/2)}{\Gamma(r + 1)} = \frac{\pi}{2^{2r+1}} \frac{(2r)!}{(r!)^2},$$

we get the following expansions:

$$K = \frac{\pi}{2} \sum_{r=0}^{\infty} \frac{(2r)!(2r)!}{2^{4r}(r!)^4} k^{2r},$$

$$E = \frac{\pi}{2} \left\{ 1 - \sum_{r=1}^{\infty} \frac{(2r - 2)!(2r)!}{2^{4r-1}(r - 1)!(r!)^3} k^{2r} \right\}.$$

We now recall the hypergeometric series

$$F(a, b, c; x) = 1 + \frac{a \cdot b}{c \cdot 1} x + \frac{a(a + 1)b(b + 1)}{c(c + 1) \cdot 1 \cdot 2} x^2 + \cdots .$$

It is not hard to verify that this series becomes $2K/\pi$ if

$$a = b = 1/2, \qquad c = 1, \qquad x = k^2,$$

and becomes $2E/\pi$ if

$$a = -1/2, \qquad b = 1/2, \qquad c = 1, \qquad x = k^2.$$

Thus,

$$K = \tfrac{\pi}{2} F(\tfrac{1}{2}, \tfrac{1}{2}, 1; k^2), \tag{1^{bis}}$$

$$E = \tfrac{\pi}{2} F(-\tfrac{1}{2}, \tfrac{1}{2}, 1, k^2), \tag{2^{bis}}$$

These expressions are suitable for $|k^2| < 1$. The known transformations of a hypergeometric series give expressions of the elliptic integrals K and E for other parts of the plane of the variable k^2. Without dwelling on these transformations, we turn to another aspect of the question, namely, connections with conformal mappings.

Suppose that we have a finite domain whose boundary is formed by circular arcs in the complex w-plane. Denote by $c_1, \ldots, c_n$ the vertices of the domain (the circular polygon), and by $\pi\delta_1, \ldots, \pi\delta_n$ the corresponding interior angles. Assume that it is required to map this polygon conformally onto the upper half-plane of the complex variable z. Moreover, assume that the image of the vertex c_n must be the point at infinity, and denote by $a_1, \ldots, a_{n-1}$

the respective points of the real axis into which the remaining vertices of the polygon are mapped.

In this case, as one proves in courses in the geometric theory of functions of a complex variable,[21] the function $w = w(z)$ satisfies the differential equation $\{w, z\} = 2I(z)$, where

$$2I(z) = \frac{1}{2} \sum_{r=1}^{n-1} \frac{1 - \delta_r^2}{(z - a_r)^2} + \sum_{r=1}^{n-1} \frac{A_r}{z - a_r}, \tag{3}$$

$A_1, \ldots, A_{n-1}$ being constants connected by the relations

$$\sum_{r=1}^{n-1} A_r = 0, \tag{4}$$

$$\sum_{r=1}^{n-1} A_r a_r + \frac{1}{2} \sum_{r=1}^{n-1} (1 - \delta_r^2) = \frac{1}{2}(1 - \delta_n^2), \tag{5}$$

and $\{w, z\}$ is the so-called Schwarzian derivative (or Schwarzian) of w with respect to z:

$$\{w, z\} = w'''/w' - \tfrac{3}{2}(w''/w')^2.$$

Thus, to find w we have a third-order differential equation. However, the solution of the problem can be reduced to the integration of a certain second-order differential equation. Indeed, we have the following statement: *every solution of the equation*

$$\{w, z\} = 2I(z) \tag{6}$$

is the ratio of two linearly independent particular integrals of the differential equation

$$d^2\Omega/dz^2 + I\Omega = 0. \tag{7}$$

For a proof we introduce in (6) a new unknown v in place of w by setting

$$w''/w' = -2v'/v. \tag{8}$$

The new equation is $v'' + Iv = 0$, i.e., v is one of the solutions of (7), say $v = \Omega_1$. On the other hand, it follows from (8) that

$$w' = C^*/v^2, \tag{8bis}$$

where C^* is a nonzero constant. We take the second solution of (7) (call it Ω_2) such that $\Omega_1\Omega_2' - \Omega_2\Omega_1' = C^*$. Then, by (8bis),

$$w' = \frac{C^*}{\Omega_1^2} = \frac{d}{dz}\frac{\Omega_2}{\Omega_1},$$

and hence

$$w = \frac{\Omega_2}{\Omega_1} + C = \frac{\Omega_2 + C\Omega_1}{\Omega_1},$$

where C is a new constant.

Thus, our statement is proved.

It is not hard to see that the converse also holds.

The case of a lune is trivial. Indeed, in this case only one of the two vertices is represented by a point lying at a finite distance, and we can take the point $z = 0$ for it. Thus,

$$2I(z) = \frac{1}{2}\frac{1 - \delta^2}{z^2} + \frac{A}{z}.$$

---

[21] See, for example, [1], Section III, Chapter 7, §6 (3rd ed.), §7 (4th ed.).

Condition (4) reduces to the equality $A = 0$, and (5) is easily seen to be satisfied automatically. Our equation (6) will have the form

$$\{w, z\} = \frac{1}{2}\frac{1 - \delta^2}{z^2},$$

while (7) can be written in the form

$$\frac{d^2\Omega}{dz^2} + \frac{1}{4}\frac{1 - \delta^2}{z^2}\Omega = 0.$$

The general integral of this equation is

$$\Omega = C_1 z^{(1+\delta)/2} + C_2 z^{(1-\delta)/2}.$$

Thus, a function realizing the desired conformal mapping is

$$w = \frac{A_1 z^{(1+\delta)/2} + A_2 z^{(1-\delta)/2}}{B_1 z^{(1+\delta)/2} + B_2 z^{(1-\delta)/2}} \quad \text{or} \quad w = \frac{A_1 z^\delta + A_2}{B_1 z^\delta + B_2}.$$

The next (and for us the most interesting) case is that in which the polygon has three vertices. The point at infinity is the image of one of them. Taking the points $z = 0$ and $z = 1$ as the images of the other two, we have that

$$2I(z) = \frac{A}{z} + \frac{B}{z - 1} + \frac{1}{2}\frac{1 - \delta_1^2}{z^2} + \frac{1}{2}\frac{1 - \delta_2^2}{(z - 1)^2}.$$

Here conditions (4) and (5) take the form

$$A + B = 0,$$
$$A \cdot 0 + B \cdot 1 = \tfrac{1}{2}(1 - \delta_3^2) - \tfrac{1}{2}(1 - \delta_1^2) - \tfrac{1}{2}(1 - \delta_2^2).$$

Thus,

$$-A = B = \tfrac{1}{2}(\delta_1^2 + \delta_2^2 - \delta_3^2 - 1),$$

and we get the equation

$$\frac{d^2\Omega}{dz^2} + \Omega\left\{\frac{1 - \delta_1^2}{4z^2} + \frac{1 - \delta_2^2}{4(z - 1)^2} - \frac{\delta_1^2 + \delta_2^2 - \delta_3^2 - 1}{4z} + \frac{\delta_1^2 + \delta_2^2 - \delta_3^2 - 1}{4(z - 1)}\right\} = 0.$$

Now let

$$\begin{aligned}
\delta_1^2 &= (1 - c)^2, \\
\delta_2^2 &= (c - a - b)^2, \\
\delta_3^2 &= (a - b)^2.
\end{aligned} \tag{9}$$

Moreover, instead of Ω we introduce the function y by the formula

$$\Omega = z^{c/2}(z - 1)^{(a+b-c+1)/2}y.$$

Then y satisfies

$$z(1 - z)y'' + [c - (a + b + 1)z]y' - aby = 0.$$

This is a differential equation satisfied by a hypergeometric function with parameters a, b, and c.

We saw above that

$$K = \tfrac{\pi}{2}F(\tfrac{1}{2}, \tfrac{1}{2}, 1; \lambda),$$

where $\lambda = k^2$. Similarly,

$$K' = \tfrac{\pi}{2}F(\tfrac{1}{2}, \tfrac{1}{2}, 1; 1 - \lambda).$$

Therefore, K is a particular integral of the equation

$$\lambda(1 - \lambda)y'' - (2\lambda - 1)y' - \tfrac{1}{4}y = 0, \tag{10}$$

and K' is another particular integral. Hence, the function

$$w \equiv \frac{\alpha i K' + \beta K}{\gamma i K' + \delta K} = \varphi(\lambda) \tag{11}$$

maps the upper half of the λ-plane onto some triangle in the w-plane. All the angles of this triangle are equal to zero, as is easily seen from (9). We can represent (11) in the form

$$\frac{\alpha \tau + \beta}{\gamma \tau + \delta} = \varphi(\lambda),$$

and, taking $\alpha = \delta = 1$ and $\beta = \gamma = 0$, we find without difficulty that the triangle in the τ-plane has its vertices at the points $\tau = 0$, $\tau = 1$, and $\tau = \infty$. We come to the right half of the domain D_2 considered in §23.

To conclude the present section we note that a differential equation equivalent to (10) was found as far back as Legendre. Namely, Legendre showed that K and K' satisfy the equation

$$\frac{d}{dk}\left(kk'^2\frac{dy}{dk}\right) = ky,$$

and this equation is equivalent to (10).

To get a differential equation satisfied by complete elliptic integrals it is not at all necessary to first expand these integrals in power series. It is simpler to use differentiation with respect to a parameter.

Indeed, from (2),

$$\frac{dE}{dk} = -\int_0^{\pi/2} \frac{k\sin^2 t}{\sqrt{1 - k^2\sin^2 t}}\,dt = \frac{E - K}{k}.$$

Further,

$$\frac{dK}{dk} = \int_0^{\pi/2} \frac{k\sin^2 t}{(1 - k^2\sin^2 t)^{3/2}}\,dt.$$

Here we set

$$\frac{k'^2}{1 - k^2\sin^2 t} = 1 - k^2\sin^2 \varphi.$$

From this,

$$\sin t = \frac{\cos\varphi}{\sqrt{1 - k^2\sin^2\varphi}}, \qquad \cos t = \frac{k'\sin\varphi}{\sqrt{1 - k^2\sin^2\varphi}},$$

and after simple computations we get

$$k\frac{dK}{dk} = \int_0^{\pi/2} \frac{k^2\cos^2\varphi\,d\varphi}{k'^2\sqrt{1 - k^2\sin^2\varphi}} = \frac{E}{k'^2} - K.$$

Elimination of E from the relations

$$\frac{dE}{dk} = \frac{E - K}{k}, \qquad k\frac{dK}{dk} = \frac{E}{k'^2} - K$$

gives us

$$\frac{d}{dk}\left(kk'^2\frac{dK}{dk}\right) = kK.$$

We propose as an exercise an analogous method of verifying that E and $E' - K'$ are solutions of the equation

$$k'^2\frac{d}{dk}\left(k\frac{dy}{dk}\right) + ky = 0.$$

The expressions found for the derivatives of the complete integrals with respect to the modulus enable us to prove the Legendre relation

$$EK' + E'K - KK' = \pi/2.$$

Indeed, by differentiation we see that the left-hand side is a constant. To determine this constant it suffices to find the limit of the left-hand side as $k \to 0$.

§45. Computation of h from a given modulus k

If the quantity $h = e^{\pi i \tau}$ is known, that it is possible to construct theta series that converge rapidly and are thus very convenient for computations. However, in practice often the modulus k is given rather than h, and then the question of finding h arises first and foremost in computations. We saw in §44 that complete elliptic integrals of the first kind can be expressed as hypergeometric series with respect to the modulus k and the complementary modulus k'. In principle these series can be used for finding K and K', and hence also $\tau = iK'/K$ and h. However, they are not well suited for computations. Weierstrass determined a very convenient path for computations under the assumption that

$$\left|\frac{1 - \sqrt{k'}}{1 + \sqrt{k'}}\right| < 1. \tag{1}$$

We always understand $\sqrt{k'}$, on which the quantity

$$l = (1 - \sqrt{k'})/(1 + \sqrt{k'}) \tag{2}$$

depends, to be the value of it whose real part is positive. In practice the most important case is that in which $0 < k < 1$. In this case it is also true that $0 < k' < 1$, and condition (1) certainly holds. As a starting point we take the formula

$$k' = \vartheta_0^2(0|\tau)/\vartheta_3^2(0|\tau),$$

which we have encountered more than once. On the basis of this formula

$$\sqrt{k'} = \vartheta_0(0|\tau)/\vartheta_3(0|\tau),$$

and hence

$$l = \frac{1 - \sqrt{k'}}{1 + \sqrt{k'}} = \frac{\vartheta_3(0|\tau) - \vartheta_0(0|\tau)}{\vartheta_3(0|\tau) + \vartheta_0(0|\tau)}.$$

Recalling that

$$\vartheta_3(0|\tau) = 1 + 2h + 2h^4 + 2h^9 + \cdots,$$
$$\vartheta_0(0|\tau) = 1 - 2h + 2h^4 - 2h^9 + \cdots,$$

we get the relation

$$l = \frac{2h + 2h^9 + 2h^{25} + \cdots}{1 + 2h^4 + 2h^{16} + \cdots}.$$

It can be rewritten in the form

$$l^4 = \vartheta_2^4(0|4\tau)/\vartheta_3^4(0|4\tau). \tag{3}$$

This differs only in notation from the equation

$$k^2 = \vartheta_2^4(0|\tau)/\vartheta_3^4(0|\tau).$$

But in §30 we saw that in view of the last equation the quantity $h = e^{\pi i \tau}$ is a regular analytic function of k^2 in the plane of the complex variable $\lambda = k^2$, cut along the real axis from the point $\lambda = 1$ to $\lambda = \infty$.

Applying this result to our equation (3), we conclude that $h^4 = e^{4\pi i \tau}$ is a regular analytic function of l^4 in the disk $|l| < 1$. Therefore, $h^4 = A_1 l^4 + A_2 l^8 + A_3 l^{12} + \cdots$, or

$$h = C_1 l + C_2 l^5 + C_3 l^9 + \cdots.$$

It is not hard to compute the first coefficients in this series. The result has the form

$$h = \tfrac{1}{2}l + 2(\tfrac{1}{2}l)^5 + 15(\tfrac{1}{2}l)^9 + 150(\tfrac{1}{2}l)^{13} + \cdots.$$

It is very convenient to use this series in practice, since the first two terms usually suffice.

§46. The arithmetico-geometric mean

Let a and b be two positive numbers with $a > b$. With their help we construct two sequences $a_1, a_2, \ldots$ and $b_1, b_2, \ldots$ by setting

$$a_1 = (a + b)/2, \qquad b_1 = \sqrt{ab},$$
$$a_2 = (a_1 + b_1)/2, \qquad b_2 = \sqrt{a_1 b_1},$$

and so on, where the root always has the arithmetic value. It is easy to prove that the quantities a_n and b_n converge as $n \to \infty$ to a common limit. This limit is called the arithmetico-geometric mean of a and b, and is denoted by $\mu(a, b)$. It was first considered by Gauss.

For a proof we note that the following inequalities hold:

$$a_n > b_n, \tag{1}$$

$$a_{n+1} < a_n, \qquad b_{n+1} > b_n. \tag{2}$$

It follows that a_n and b_n have limits:

$$\alpha = \lim_{n \to \infty} a_n, \qquad \beta = \lim_{n \to \infty} b_n.$$

The existence of the limits α and β implies that

$$\alpha = \lim_{n \to \infty} a_{n+1} = \lim_{n \to \infty} \frac{a_n + b_n}{2} = \frac{\alpha + \beta}{2},$$

or $\alpha = \beta$.

We now occupy ourselves with finding $\mu(a, b)$. With this goal we take the basic relation expressing Gauss's transformation:

$$\operatorname{sn}\left(u; \frac{2\sqrt{k}}{1+k}\right) = \frac{(1+k)\operatorname{sn}(u/(1+k); k)}{1 + k\,\operatorname{sn}^2(u/(1+k); k)}.$$

Setting

$$\operatorname{sn}\left(u; \frac{2\sqrt{k}}{1+k}\right) = \sin \varphi, \qquad \operatorname{sn}\left(\frac{u}{1+k}; k\right) = \sin \psi,$$

so that

$$u = \int_0^\varphi \frac{dt}{\sqrt{1 - (4k/(1+k)^2)\sin^2 t}}, \qquad \frac{u}{1+k} = \int_0^\psi \frac{dt}{\sqrt{1 - k^2 \sin^2 t}},$$

we can write

$$\frac{1}{1+k} \int_0^\varphi \frac{dt}{\sqrt{1 - (4k/(1+k)^2)\sin^2 t}} = \int_0^\psi \frac{dt}{\sqrt{1 - k^2 \sin^2 t}}, \tag{4}$$

where φ and ψ are connected by the following relation, which is a consequence of (3):

$$\sin \varphi = \frac{(1+k)\sin \psi}{1 + k \sin^2 \psi}.$$

Let us consider the special case when $\psi = \pi/2$, and hence $\varphi = \pi/2$. Then (4) takes the form

$$\frac{1}{1+k} \int_0^{\pi/2} \frac{dt}{\sqrt{1 - (4k/(1+k)^2)\sin^2 t}} = \int_0^{\pi/2} \frac{dt}{\sqrt{1 - k^2 \sin^2 t}}. \tag{5}$$

Let $k = (a_n - b_n)/(a_n + b_n)$, so that

$$1 + k = \frac{a_n}{a_{n+1}}, \qquad k^2 = 1 - \frac{b_{n+1}^2}{a_{n+1}^2}, \qquad \frac{4k}{(1+k)^2} = 1 - \frac{b_n^2}{a_n^2},$$

and (5) can be rewritten as follows:

$$\int_0^{\pi/2} \frac{dt}{\sqrt{a_n^2 \cos^2 t + b_n^2 \sin^2 t}} = \int_0^{\pi/2} \frac{dt}{\sqrt{a_{n+1}^2 \cos^2 t + b_{n+1}^2 \sin^2 t}}.$$

We see that the quantity

$$\int_0^{\pi/2} \frac{dt}{\sqrt{a_n^2 \cos^2 t + b_n^2 \sin^2 t}}$$

does not depend on n, and is thus equal to its limit as $n \to \infty$. Consequently,

$$\int_0^{\pi/2} \frac{dt}{\sqrt{a^2 \cos^2 t + b^2 \sin^2 t}} = \frac{1}{\mu(a,b)} \int_0^{\pi/2} \frac{dt}{\sqrt{\cos^2 t + \sin^2 t}},$$

i.e.,

$$\mu(a,b) = \frac{\pi}{2 \int_0^{\pi/2} dt / \sqrt{a^2 \cos^2 t + b^2 \sin^2 t}}.$$

CHAPTER 8

Some Conformal Mappings

§47. A conformal mapping of a rectangle onto a half-plane

Suppose that a rectangle with vertices at the points $u = a$, $a + bi$, $-a + bi$, $-a$ is given in the u-plane, where a and b are positive numbers. It is required to map this rectangle conformally onto the upper half of the z-plane. As is known from the theory of functions of a complex variable, the desired mapping function (for the time being we are concerned with finite points) is continuous up to the boundary.

If we denote by c_i $(i = 1, 2, 3, 4)$ the points of the real axis that are the images of the vertices of the rectangle, then by the well-known Schwarz-Christoffel formula

$$C'u + C'' = \int_0^z \frac{dx}{\sqrt{(x - c_1)(x - c_2)(x - c_3)(x - c_4)}}.$$

On the basis of general theorems, the mapping function is completely determined if the images of three boundary points of the rectangle are given. We require that the points $z = -1, 0$, and 1 correspond to the points $u = -a, 0$, and a (Figure 11). These requirements determine three of the constants, namely, we get that $C'' = 0$, $c_3 = -1$, and $c_4 = 1$. Thus

$$C'u = \int_0^z \frac{dx}{\sqrt{(x^2 - 1)(x - c_1)(x - c_2)}}. \tag{1}$$

According to the Riemann-Schwarz symmetry principle, the function u can be continued analytically across the segment $[-1, 1]$ of the real axis in the z-plane. We obtain the rectangle that is symmetric to the given one with respect to the real axis (the lower rectangle), and (1) gives a mapping of the lower half-plane onto this rectangle. The same mapping is obtained if u is replaced by $-u$ and z by $-z$ in (1). But if

$$-C'u = \int_0^{-z} \frac{dx}{\sqrt{(x^2 - 1)(x - c_1)(x - c_2)}},$$

119

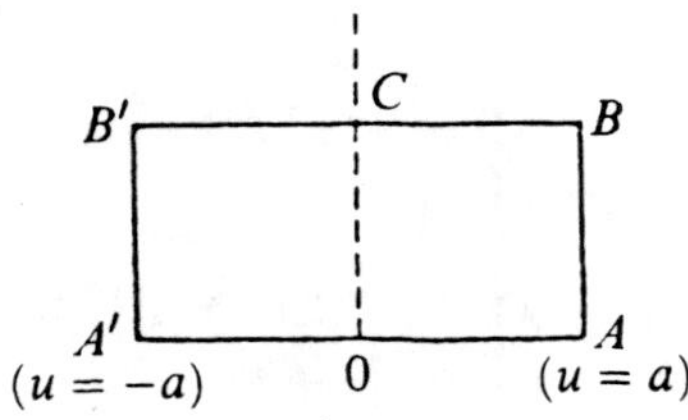

FIGURE 11

then

$$C'u = \int_0^z \frac{dx}{\sqrt{(x^2 - 1)(x + c_1)(x + c_2)}}. \tag{2}$$

Because of the uniqueness of the mapping function under the assumed correspondence of the three boundary points, the function (1) must be identical with (2), which implies that

$$(x - c_1)(x - c_2) = (x + c_1)(x + c_2),$$

and hence $c_2 = -c_1$. Thus,

$$(x - c_1)(x - c_2) = x^2 - c_1^2,$$

and replacing c_1 by $1/k$, where k can be assumed to be positive and less than 1, we represent the mapping function in the form

$$u = \frac{1}{C} \int_0^z \frac{dx}{\sqrt{(1 - x^2)(1 - k^2x^2)}}. \tag{3}$$

We now have only the two parameters C and k. To determine them we have the following equations:

$$a = \frac{1}{C} \int_0^1 \frac{dx}{\sqrt{(1 - x^2)(1 - k^2x^2)}}, \tag{4'}$$

$$bi = \frac{i}{C} \int_1^{1/k} \frac{dx}{\sqrt{(x^2 - 1)(1 - k^2x^2)}}. \tag{4''}$$

Dividing the second by the first, we arrive at an equation for k:

$$\frac{b}{a} = \int_1^{1/k} \frac{dx}{\sqrt{(x^2 - 1)(1 - k^2x^2)}} \bigg/ \int_0^1 \frac{dx}{\sqrt{(1 - x^2)(1 - k^2x^2)}}. \tag{4}$$

Once k is determined, C can be found from (4') (or (4'')).

Let us study equation (4). With this goal we transform the integral

$$\int_1^{1/k} \frac{dx}{\sqrt{(x^2-1)(1-k^2x^2)}},$$

setting

$$k^2x^2 + k'^2y^2 = 1 \tag{5}$$

and assuming that $0 \le y \le 1$ when $1/k \ge x \ge 1$. By (5),

$$k'\,dy/\sqrt{1-k'^2y^2} = -k\,dx/\sqrt{1-k^2x^2}.$$

But since, moreover, $k\sqrt{x^2-1} = k'\sqrt{1-y^2}$, it follows that

$$\int_1^{1/k} \frac{dx}{\sqrt{(x^2-1)(1-k^2x^2)}} = \int_0^1 \frac{dy}{\sqrt{(1-y^2)(1-k'^2y^2)}}.$$

Thus, (4) takes the form

$$\frac{b}{a} = \int_0^1 \frac{dx}{\sqrt{(1-x^2)(1-k'^2x^2)}} \Big/ \int_0^1 \frac{dx}{\sqrt{(1-x^2)(1-k^2x^2)}}. \tag{4bis}$$

On the right-hand side we have the ratio of the complete elliptic integrals of the first kind K' and K for the moduli k' and k. It is easy to see that the right-hand side varies monotonically from ∞ to 0 as k increases from 0 to 1. This makes it clear that for an arbitrary value of the ratio b/a there exists a k in $(0,1)$ such that (4) holds. The mapping function is

$$u = \frac{a}{K} \int_0^z \frac{dx}{\sqrt{(1-x^2)(1-k^2x^2)}}. \tag{3bis}$$

The same result could have been obtained by starting from the function

$$z = \operatorname{sn}(Ku/a; k)$$

and studying its real values. We show this.

Suppose that the point u moves in the positive direction along the boundary of our rectangle, starting from $u = 0$. The point $z = 0$ corresponds to this position. The variable z increases from $z = 0$ to $z = 1$ as u increases from 0 to a. We proceed to the side AB of our rectangle. On this side $u = a + iv$, where v varies from 0 to b. But since

$$\operatorname{sn}(K + iw; k) = \frac{\operatorname{cn}(iw; k)}{\operatorname{dn}(iw; k)} = \frac{1}{\operatorname{dn}(w; k')},$$

it follows that on the side AB

$$z = 1/\operatorname{dn}(Kv/a; k'),$$

where v varies from 0 to b, and hence z increases from 1 to $1/\mathrm{dn}(K';k') = 1/k$. We proceed to the side BC, on which $u = ib + v$ and v varies from a to 0. Since

$$z = \mathrm{sn}\left(iK' + \frac{Kv}{a};k\right) = \frac{1}{k\,\mathrm{sn}(Kv/a;k)},$$

z varies from $1/k$ to ∞.

We see that the right half of the real axis in the z-plane corresponds to the right half $OABC$ of the boundary of the rectangle. The fact that the left half of the real axis corresponds to the left half of the boundary of the rectangle no longer requires a special proof.

Since the function $z = \mathrm{sn}(Ku/a;k)$ is regular interior to the rectangle and since it maps the boundary of this rectangle in a one-to-one fashion onto the real axis, the function under consideration maps the rectangle onto the half-plane.

The path (which is independent of the theory of elliptic functions) by which we arrived at the mapping function (3bis) is of interest because it enables us for the normal case $(0 < k < 1)$ to introduce the basic elliptic function of Jacobi and to discover its most important properties.

For simplicity let use assume that $a = K$, and hence $b = K'$. We have the function

$$u = f(z) \equiv \int_0^z \frac{dx}{\sqrt{(1 - x^2)(1 - k^2 x^2)}}, \tag{6}$$

which maps the upper half of the z-plane conformally onto the rectangle R in the u-plane (Figures 12 and 13). The function u will now be continued analytically according to the Riemann-Schwarz symmetry principle. First of all, we can continue our function across the segment $[-1, 1]$ to the lower half of the z-plane. In the u-plane we get the rectangle R^{-1}. A further continuation is carried out across the segment $[-1/k, -1]$, to which the side IV of the rectangle R^{-1} corresponds. We again get the upper half of the z-plane, and we get the rectangle R^{-1}_{-1} in the u-plane. Then we continue again across $[-1, 1]$, and this leads us to the rectangle R_{-1}, which is a mapping of the lower half-plane. Going on with this process, we keep getting new rectangles, which in the limit cover the whole u-plane. Each unshaded rectangle is a mapping of the upper half of the z-plane and each shaded rectangle a mapping of the lower half.

Since the rectangles in the u-plane do not overlap, a completely determined value of z corresponds to each value of u, i.e., z is a single-valued function of u. We know in advance that this function is analytic. The points $iK' + 2mK + 2inK'$ with m and n integers are the only poles of the function $z = g(u)$.

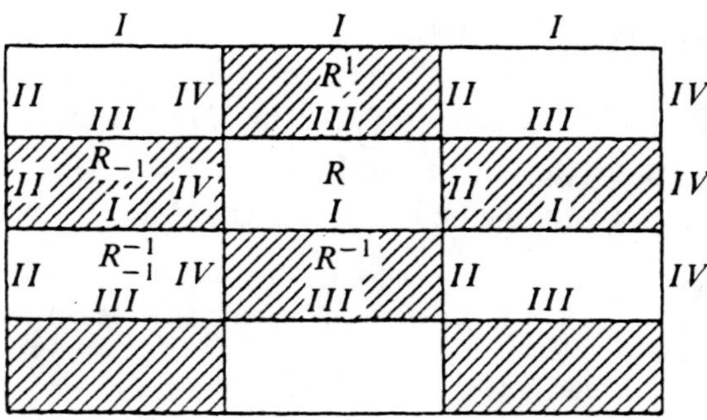

FIGURE 12

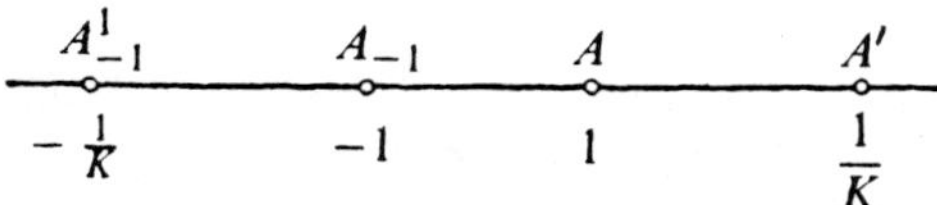

FIGURE 13

It is also easy to see that these points are simple poles on the basis of the properties of the mapping function. Indeed, take the rectangle consisting of R and R^1. It is mapped onto the whole z-plane, cut along the segment $[-1/k, 1/k]$. The point at infinity is an interior point of this domain in the z-plane. Corresponding to it is the point $u = iK'$ on the segment III. If the function $z = g(u)$ had a pole of higher order at this point, then corresponding to a simple circuit about the point $u = iK'$ would be a multiple circuit about the point $z = \infty$ in the z-plane, and this is impossible in view of the one-to-oneness of the conformal mapping.

Accordingly, our assertion is proved.

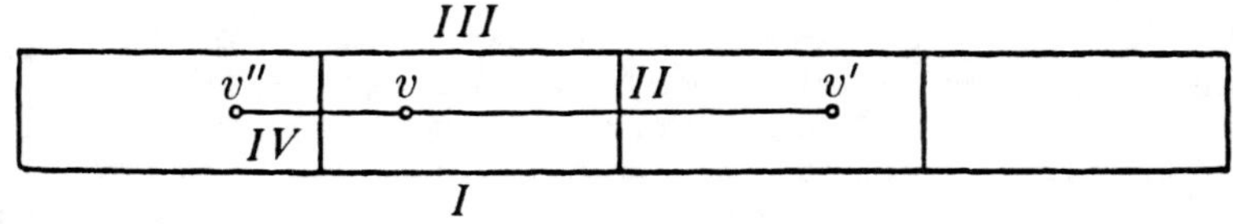

FIGURE 14

The periodity of the function $z = g(u)$ can be proved in a very simple manner. Indeed, taking an arbitrary point v in the rectangle R (Figure 14), we find a point v' symmetric to v with respect to the side II, and then a point v'' symmetric to v with respect to the side IV. The function $g(u)$

has the same values at the points v' and v'': $g(v') = g(v'')$. At the same time it is easy to see that $v' - v''$ is equal to double the length of I, i.e., is equal to $4K$. It can be proved similarly that $2iK'$ is the second period.

Thus, the basic properties of the function $z = \operatorname{sn}(u; k)$, i.e., the upper limit as a function of the value of the integral, have been verified.

We consider also the domain onto which the function $z = \operatorname{sn}(u; k)$ maps not just one rectangle, say R, and not just a pair of adjacent rectangles, but the whole plane of the complex variable u.

To each rectangle of our family there corresponds a half-plane: the upper half-plane corresponds to an unshaded rectangle, and the lower half-plane corresponds to a shaded rectangle (Figure 15). Since there are infinitely many rectangles, we prepare an infinite set of half-planes (shaded and unshaded) and arrange them in such a way that the unshaded and shaded ones adjoin each other along the real axis, and points of all the planes having the same coordinates lie one over the other.

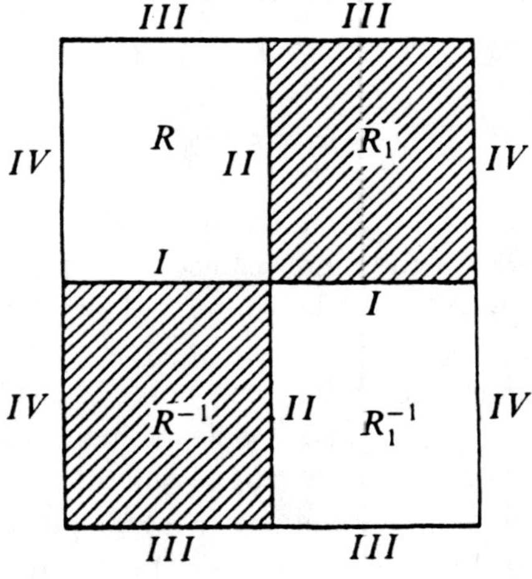

FIGURE 15

Since we want to get the full u-plane, we must sew the rectangles together along certain sides. The segments of the boundaries of the half-planes corresponding to the sewn rectangles also have to be sewn. As a result we get over the z-plane an infinite-sheeted Riemann surface that is the image of our u-plane obtained by means of the function $z = \operatorname{sn}(u; k)$.

To study elliptic functions with periods $4K$ and $2iK'$ it suffices to work with four rectangles of the family considered: R, R^{-1}, R_1^{-1}, and R_1 (Figures 15 and 16). Corresponding to the pair of rectangles R, R^{-1} is a pair of half-planes denoted by the same letters. Sewing the rectangles R and R^{-1} together along I, we must sew the indicated half-planes together along the segment $[-1, 1]$. We obtain a plane with cuts along the semi-axes $(-\infty, -1)$ and $(1, \infty)$. Sewing R_1 and R_1^{-1} together along I, we get a second plane with cuts along the same semi-axes $(-\infty, -1)$ and $(1, \infty)$.

We now sew the double rectangles together along II, and also along IV. The planes are subjected to an analogous sewing operation. Further, to get continuity of the correspondence

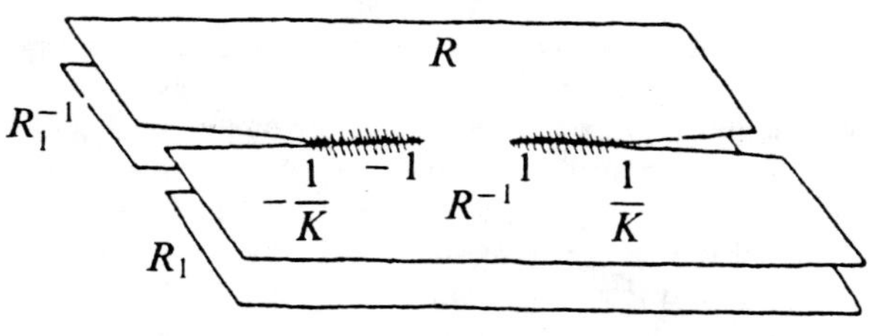

FIGURE 16

the upper edge of the cut $(1, 1/k)$ in the upper sheet must be sewn together with the lower edge of the corresponding cut in the lower sheet.

The same applies to the cut $(-1/k, -1)$. As a result we get a two-sheeted Riemann surface with lines of transition along the segments $(-1/k, -1)$ and $(1, 1/k)$ and with cuts along $(-\infty, -1/k)$ and $(1/k, \infty)$.

A piece of a cylindrical surface is obtained after sewing together the rectangles (Figure 17). The function $z = g(u)$ takes the same values at opposite points of the sides III of the double rectangles in the z-plane. These sides must therefore be sewn together. Passing to the cylinder, we must deform it and sew together its boundaries. We obtain a torus (Figure 18).

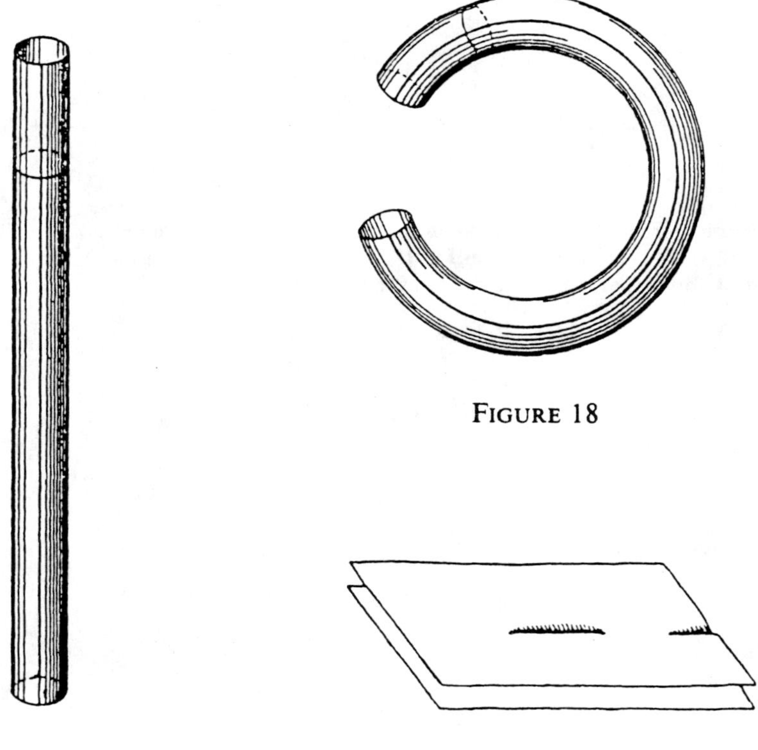

FIGURE 18

FIGURE 17

FIGURE 19

The sheets of the Riemann surface must be subjected to a corresponding sewing operation. Half-sheets of one and the same plane must be sewn together here.

As a result we obtain the two-sheeted Riemann surface pictured in Figure 19. This surface no longer has cuts. It has four branch points and two transition lines.

Accordingly, we have three images: the two-sheeted Riemann surface just constructed, the rectangle consisting of the four rectangles R, R_1, R^{-1}, and R_1^{-1}, with the corresponding points of opposite sides of this rectangle identified, and, finally, the torus.

The rectangle is mapped onto the Riemann surface conformally, and the torus is topologically equivalent to the rectangle. Thus all three images are topologically equivalent.[22]

However, it does not hurt to show that the torus can be mapped onto the rectangle not only in a one-to-one and continuous fashion, as we did above by means of deformation and sewing together of opposite sides, but also conformally.

Assume that the torus was obtained by rotating about the Z-axis a circle lying in the OXZ plane and having the equation

$$(X - R)^2 + Z^2 = \rho^2.$$

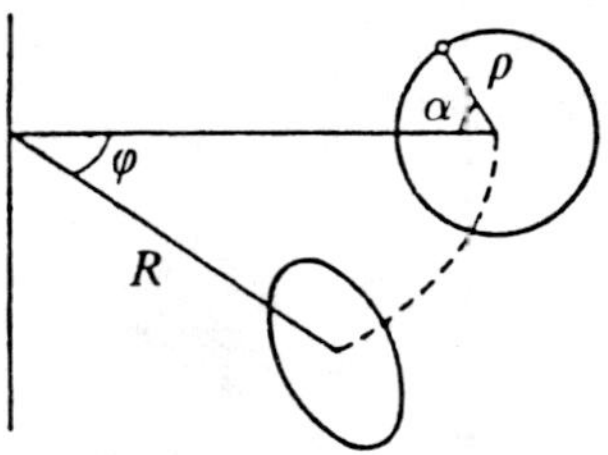

$$\textbf{Figure 20}$$

To determine the position of a point on the torus we can use the two angles $\alpha, \varphi \in [0, 2\pi)$, the sense of which is easily seen from Figure 20, and from the expressions we give for the Cartesian coordinates of a point on the torus in terms of these angles:

$$X = (R - \rho \cos \alpha) \cos \varphi,$$
$$Y = (R - \rho \cos \alpha) \sin \varphi,$$
$$Z = \rho \sin \alpha.$$

The element of arclength is

$$ds = \sqrt{dX^2 + dY^2 + dZ^2} = \sqrt{(R - \rho \cos \alpha)^2 d\varphi^2 + \rho^2 d\alpha^2}.$$

This expression can be reduced to the form

$$ds = (R - \rho \cos \alpha)\sqrt{d\xi^2 + d\eta^2}$$

by setting

$$\xi = \varphi, \qquad \eta = \int_0^a \frac{\rho \, d\alpha}{R - \rho \cos \alpha}. \tag{7}$$

---

[22]The Riemann surface can be assumed to be constructed by means of Neumann spheres instead of planar sheets.

As φ and α vary from 0 to 2π, the domain of variation of the point (ξ, η) in the (ξ, η)-plane is the rectangle with sides the segment $[0, 2\pi]$ of the ξ-axis and the segment $[0, l]$ of the η-axis, where

$$l = \int_0^{2\pi} \frac{\rho \, d\alpha}{R - \rho \cos \alpha}.$$

We hold to the convention indicated above about identifying opposite sides of the rectangle.

By means of (7) this rectangle is mapped onto the torus not only in a one-to-one and continuous manner, but also conformally, since in the expression for the arclength differential the coefficients of $d\xi^2$ and $d\eta^2$ are the same, while the coefficient of $d\xi \, d\eta$ is equal to zero.

A rectangle with each pair of opposite sides identified is none other than a period rectangle. A class of elliptic functions belongs to it. Each function in this class can be regarded as a single-valued analytic function on the torus (with only finitely many poles as the only singular points). Corresponding to this class of functions (on the rectangle and on the torus) is a class of single-valued analytic functions on the two-sheeted Riemann surface, namely, all rational functions of z and $\sqrt{(1 - z^2)(1 - k^2 z^2)}$.

All these facts are important not only in the consideration of elliptic functions and elliptic integrals, but also in the consideration of the algebraic function $w(z)$ determined by

$$w^2 = f(z), \tag{8}$$

where $f(z)$ is a fourth-degree polynomial without multiple roots.

If instead of (8) we take an equation with a polynomial $f(z)$ of higher degree (but without multiple roots), then it is also possible to introduce a two-sheeted Riemann surface on which every rational function $R(z, w)$ is single-valued. On this surface it is also possible to study the integrals $\int R(z, w) \, dz$, which generalize elliptic integrals. However, in general the subsequent constructions turn out not to be as simple as in the "elliptic" case we considered. Nevertheless, uniformization is possible, but with the help of functions that are not elliptic but automorphic. In this book we cannot concern ourselves with these questions, nor with the analogous questions arising when (8) is replaced by a general algebraic equation $F(z, w) = 0.$[23]

§48. A conformal mapping of a doubly connected polygonal domain onto a circular annulus

It is well known that arbitrary simply connected domains (whose boundaries have at least two points each) can be mapped conformally one onto another. The situation is different with doubly connected domains. For example, take the two circular annuli

$$g: r_1 \le |z| \le r_2 \qquad (r_2 > r_1 > 0),$$
$$G: R_1 \le |Z| \le R_2 \qquad (R_2 > R_1 > 0).$$

A conformal mapping of one of these annuli onto the other can be obtained by setting $Z = Az$ or $Z = B/z$, where A and B are constants. In the first case we have that $R_2 = |A|r_2$ and $R_1 = |A|r_1$, which implies that

---

[23] With the exception of Abel's theorem, which is relevant here and to which §54 is devoted.

$R_1/R_2 = r_1/r_2$. In the second case $R_2 = |B|/r_1$ and $R_1 = |B|/r_2$, and hence again $R_1/R_2 = r_1/r_2$.

It is not hard to see that these are the only one-to-one conformal mappings of g onto G. Indeed, by successive application of the Riemann-Schwarz symmetry principle, a function $Z = Z(z)$ giving the required mapping can be continued to the whole z-plane with the points $z = 0$ and $z = \infty$ removed, and this continuation of $Z(z)$ maps the twice punctured z-plane onto the same domain in the Z-plane. One of the points $z = 0$ or $z = \infty$ will be a removable singularity, and hence a root for the function $Z(z)$, while the other will be a root for $1/Z(z)$. To a simple circuit about each of the points $z = 0$ and $z = \infty$ there corresponds a simple circuit about the images of these points in the Z-plane (because the mapping is one-to-one); therefore, one of the points $z = 0$ and $z = \infty$ is a simple root and the only root, while the other is a simple pole and the only pole of the function $Z(z)$. This implies our assertion about the absence of mappings other than those considered.

We see that one circular annulus is mapped in a one-to-one and conformal fashion onto another if and only if the ratios of the radii of the circles bounding the annuli are the same.

A fortiori, an arbitrary doubly connected domain cannot be mapped in the required way onto an arbitrary circular annulus. However, it can be proved that for each doubly connected domain there exists a circular annulus onto which it can be mapped in a one-to-one and conformal manner. The ratio of the radii of the circles bounding this annulus is a parameter characterizing the doubly connected domain in the sense that only doubly connected domains with the same value of this parameter can be mapped conformally onto each other.

We consider doubly connected domains with polygons as boundaries, and we find the form of the functions mapping circular annuli onto such domains in a one-to-one and conformal manner. The formulas we derive[24] amount to a distinctive generalization of the Schwarz-Christoffel formulas used to implement a conformal mapping of a disk onto a simply connected polygonal domain. The difficulty of determining the constants in the Schwarz-Christoffel formulas is a great deficiency of these formulas from the practical point of view. Of course, our generalizations of the Schwarz-Christoffel formulas also suffer from this deficiency, and the parameter of the domain (i.e., the ratio of the radii of the circles bounding the annulus) is a further addition to the number of unknown constants here.

Let us consider in detail the case when the point at infinity is an interior point of our polynomial domain S (in the plane of the complex variables z). In other words, S is a domain outside two disjoint polygons, which we denote by A_0 and A_1.

---

[24] They were obtained by the author in 1928 (see [21], pp. 223–231).

The interior angles of the domain at the vertices of the polygons A_0 and A_1 are denoted by $\pi\alpha_1, \ldots, \pi\alpha_n$. Let 1 and h be the radii of the circles C_0 and C_1 bounding the annulus G (in the w-plane), with $0 < h < 1$ unknown in advance and to be determined.

Finally, let $a_1, \ldots, a_n$ denote the points of C_0 and C_1 that are the inverse images of the vertices of the domain.

Assume further that the inverse image of the point at infinity in S is a point $w = c$ with $h < c < 1$ on the positive half of the real axis. It is easy to see that this can be assumed and is not essential. The desired function $z = z(w)$ has a first-order pole at the point $w = c$, and it is regular at the other points of G and is continuous up to the boundary. Therefore, we can use the Riemann-Schwarz symmetry principle and continue $z(w)$ outside G. In the first place we can perform a mirror reflection in the z-plane with respect to some side of the polygon A_0 and a mirror reflection in the w-plane with respect to the circle C_0. Thus, $z(w)$ is continued to the annulus G_{-1} bounded by C_0 and the circle C_{-1} given by $|w| = 1/h$. The function $z = z(w)$ has a simple pole $w = 1/c$ in the annulus G_{-1}. We obtain the annuli $G_1, G_{-2},$ and G_2 upon further mirror reflections. An even number of successive mirror reflections in the z-plane is easily seen to be equivalent to some translation and some rotation in the z-plane. The corresponding transformation in the w-plane is given by $w_1 = h^{2k} w$. Since a translation and rotation in the z-plane can be expressed by the formula $z_1 = az + b$, where a and b are constants, the desired function $z = z(w)$ must satisfy the relation

$$z(h^{2k} w) = az(w) + b.$$

From this,

$$\frac{d}{dw} z(h^{2k} w) = a \frac{d}{dw} z(w),$$

and, further,

$$\frac{d^2 z(h^{2k} w)/dw^2}{dz(h^{2k} w)/dw} = \frac{d^2 z(w)/dw^2}{dz(w)/dw}.$$

This equality can be rewritten in the form

$$h^{2k} \frac{z''(h^{2k} w)}{z'(h^{2k} w)} = \frac{z''(w)}{z'(w)},$$

and we see that the function

$$\Phi(w) = w z''(w)/z'(w)$$

satisfies the relation

$$\Phi(h^{2k} w) = \Phi(w) \qquad (k = 0, \pm 1, \pm 2, \ldots).$$

However, it suffices to write this relation for $k = 1$:

$$\Phi(h^2 w) = \Phi(w). \tag{1}$$

Take an arbitrary positive number ω and, after determining a purely imaginary number ω' such that $h = e^{\pi i \omega'/\omega}$, let

$$\Phi(w) = \varphi\left(\frac{\omega}{\pi i} \ln w\right).$$

We then find by virtue of (1) that $\varphi(u)$ satisfies the relation $\varphi(u + 2\omega') = \varphi(u)$.

Consider now that $\Phi(w)$ does not change when the point w traverses a closed contour that lies in one of the annuli G_k ($G_0 = G$) and encircles the point $w = 0$. Since the argument of w increases by 2π during this circuit, and hence the quantity $u = (\omega/\pi i)\ln w$ changes by 2ω, it follows that

$$\varphi(u + 2\omega) = \varphi(u). \tag{2}$$

We see that $\varphi(u)$ is a doubly periodic function with periods 2ω and $2\omega'$. To construct it we must investigate the singular points of this function in some period rectangle. This reduces to a consideration of the singular points of $\Phi(w)$ in some circular annulus, which could be taken to be the annulus with C_{-1} and C as boundaries. However, it is better to take the annulus Q bounded by the circles $|w| = \varepsilon h^{-1}$ and $|w| = \varepsilon h$, where ε is less than 1 but close to 1. On the boundaries of this annulus $\Phi(w)$ is regular, and inside the annulus its only singular points are $w = a_k, c$, and $1/c$ ($k = 1, \ldots, n$); further, as is easy to see, for these points we have the decompositions

$$z = L'' + L'(w - a_k)^{\alpha_k} + (w - a_k)^{\alpha_k + 1}\mathfrak{P}(w - a_k),$$

$$z = \frac{L'}{w - c} + \mathfrak{P}(w - c),$$

$$z = \frac{L'}{cw - 1} + \mathfrak{P}\left(w - \frac{1}{c}\right),$$

where $\mathfrak{P}$ is the usual notation of a power series, and L'' and L' are constants, different in each formula, with $L' \neq 0$. The corresponding expansions of $\Phi(w)$ have the form

$$\Phi(w) = \frac{a_k(\alpha_k - 1)}{w - a_k} + \cdots,$$

$$\Phi(w) = -\frac{2c}{w - c} + \cdots,$$

$$\Phi(w) = -\frac{2}{cw - 1} + \cdots.$$

With the help of these formulas we find that for the points under consideration

$$\varphi(u) = \frac{\omega}{\pi i} \frac{\alpha_k - 1}{u - (\omega/\pi i)\ln a_k} + \cdots,$$

$$\varphi(u) = -\frac{\omega}{\pi i} \frac{2}{u - (\omega/\pi i)\ln c} + \cdots,$$

$$\varphi(u) = -\frac{\omega}{\pi i} \frac{2}{u + (\omega/\pi i)\ln c} + \cdots.$$

Thus, $\varphi(u)$ has only poles, and they are simple. The sum of the residues is equal to zero, since

$$\sum_{k=1}^{n}(\alpha_k - 1) = 4$$

in view of the theorem on the sum of the exterior angles of a polygon. We see that $\varphi(u)$ is an elliptic function, and on the basis of the general theorem in §14

$$\varphi(u) = \frac{\omega}{\pi i} \sum_{k=1}^{n}(\alpha_k - 1)\zeta\left(u - \frac{\omega}{\pi i}\ln a_k\right)$$
$$- \frac{2\omega}{\pi i}\zeta\left(u - \frac{\omega}{\pi i}\ln c\right) - \frac{2\omega}{\pi i}\zeta\left(u + \frac{\omega}{\pi i}\ln c\right) + L,$$

where L is a constant. Integrating and passing from logarithms to numbers, we get that

$$z'(w) = \mu w^{\lambda} \frac{\prod_{k=1}^{n}[\vartheta_1(\frac{\ln w - \ln a_k}{2\pi i})]^{\alpha_k - 1}}{\vartheta_1^2(\frac{\ln w - \ln c}{2\pi i})\vartheta_1^2(\frac{\ln w + \ln c}{2\pi i})} \tag{3}$$

where μ and λ are again constants.

Now let w traverse a closed contour encircling the point $w = 0$ and lying in one of the annuli G_k. The left-hand side is unchanged, but on the right-hand side the factor

$$e^{2\pi i\lambda}e^{\pi i\left[\sum_{k=1}^{n}(\alpha_k-1)-4\right]} = e^{2\pi i\lambda}$$

appears. From this it is clear that λ must be an integer.

If we subject the domain S to a continuous deformation, then both h and $z(w)$ will change in a continuous way. Therefore, λ will also change in a continuous way. Being an integer, λ must remain unchanged. To find λ we consider a deformation of S under which the polygon A_1 contracts to a point without a change in the angles, so that in the limit we obtain a mapping of the domain outside A_0 onto a disk; as is easy to see, the parameter h is equal to zero in the limit. For the function mapping the limit domain we have that

$$z_0'(w) = \mu_0 \prod_{k=1}^{m} (w - \tilde{a}_k)^{\alpha_k - 1} \frac{1}{(w - c)^2 (w - 1/c)^2} \qquad (3')$$

if we assume that the first m vertices of S are the vertices of A_0. At the same time the function (3) becomes, in the limit,

$$\mu w^\lambda \prod_{k=1}^{m} (w - a_k)^{\alpha_k - 1} w^{\sum_{m+1}^{n} (\alpha_j - 1)} \frac{1}{(w - c)^2 (w - 1/c)^2}, \qquad (3'')$$

where the parameters μ, a_k, and c can have values different from those in (3). However, it can be assumed that the parameter c is the same in $(3')$ and $(3'')$. Since

$$\sum_{j=m+1}^{n} (\alpha_j - 1) = 2,$$

we conclude from a comparison of $(3')$ and $(3'')$ that $\lambda + 2 = 0$, and hence (3) must have the form

$$z'(w) = \frac{\mu}{w^2} \frac{\prod_{k=1}^{n} [\vartheta_1(\frac{\ln w - \ln a_k}{2\pi i})]^{\alpha_k - 1}}{\vartheta_1^2(\frac{\ln w - \ln c}{2\pi i}) \vartheta_1^2(\frac{\ln w + \ln c}{2\pi i})}.$$

From this we get

$$c_1 z + c_2 = \int \frac{\prod_{k=1}^{n} [\vartheta_1(\frac{\ln w - \ln a_k}{2\pi i})]^{\alpha_k - 1}}{\vartheta_1^2(\frac{\ln w - \ln c}{2\pi i}) \vartheta_1^2(\frac{\ln w + \ln c}{2\pi i})} \frac{dw}{w^2}.$$

We remark further that there is a relation between the parameters to be determined. To obtain it we note that z does not change when w completes a circuit about the point $w = c$. Finding the residue of the integrand with respect to the point $w = c$ and equating it to zero, we get the desired relation

$$\sum_{k=1}^{n} (\alpha_k - 1) \frac{\vartheta_1'(\frac{\ln c - \ln a_k}{2\pi i})}{\vartheta_1(\frac{\ln c - \ln a_k}{2\pi i})} - 2 \frac{\vartheta_1'(\frac{\ln c}{\pi i})}{\vartheta_1(\frac{\ln c}{\pi i})} = 2\pi i.$$

The case when the domain is finite is handled in a completely analogous way. In this case, denoting the interior angles of the domain as before by $\pi \alpha_k$, we find the following formula for the mapping function:

$$c_1 z + c_2 = \int \prod_{k=1}^{n} \left[\vartheta_1 \left(\frac{\ln w - \ln a_k}{2\pi i} \right) \right]^{\alpha_k - 1} \frac{dw}{w^2}.$$

§49. Examples of conformal mappings

In this section we consider examples of conformal mappings of doubly connected polygonal domains. In the first two examples we use the formulas derived in §48, although in the second the mapping function can also be obtained without resorting to the general theory.

EXAMPLE 1. Map the domain bounded by two concentric regular n-gons with corresponding vertices on a single ray emanating from the center, onto an annulus. Let 1 and h be the radii of the circles bounding the annulus $(0 < h < 1)$, and assume that 1 is the image of the vertex A_1 of the outer polygon (Figure 21).

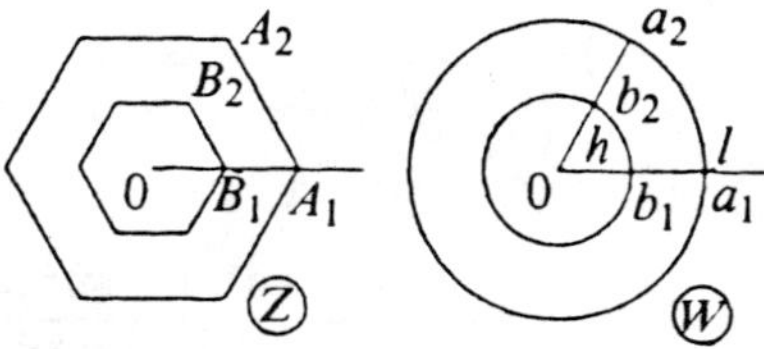

FIGURE 21

Then it follows from symmetry considerations that the points $a_r = e^{2(r-1)\pi i/n}$ and $b_r = ha_r$ $(r = 1,\ldots,n)$ in the plane of the annulus correspond to the vertices A_r and B_r.

We now introduce a purely imaginary number τ $(\tau/i > 0)$ such that $h = e^{\pi i \tau}$.

Since the interior angles of the polygonal domain are equal, respectively, to $\pi\alpha_r = (1 - 2/n)\pi$ at the vertex A_r and $\pi\beta_r = (1 + 2/n)\pi$ at the vertex B_r, we get the following expression for the mapping function by using the general formula:

$$z - 1 = C' \int_1^w \prod_{r=1}^n \left[\frac{\vartheta_1\left(\frac{\ln u - \ln ha_r}{2\pi i}\right)}{\vartheta_1\left(\frac{\ln u - \ln a_r}{2\pi i}\right)} \right]^{2/n} \frac{du}{u^2}, \tag{1}$$

where $\vartheta_1(v) = \vartheta_1(v|\tau)$. But

$$\vartheta_1\left(\frac{\ln u - \ln ha_r}{2\pi i}\right) = \vartheta_1\left(\frac{1}{2\pi i}\ln\frac{u}{a_r} - \frac{\tau}{2}\right)$$

$$= \text{const}\, u^{1/2}\vartheta_0\left(\frac{1}{2\pi i}\ln\frac{u}{a_r}\right);$$

therefore,

$$\frac{\vartheta_1((\ln u - \ln ha_r)/2\pi i)}{\vartheta_1((\ln u - \ln a_r)/2\pi i)}$$

$$= \operatorname{const} u^{1/2} \frac{\vartheta_0(\ln(u/a_r)/2\pi i)}{\vartheta_1(\ln(u/a_r)/2\pi i)}$$

$$= \operatorname{const} \frac{u^{1/2}}{\operatorname{sn}((K/\pi i)\ln(u/a_r); k)}.$$

Using the last equality, we can give the mapping function (1) the form
$(^{25})$

$$z - 1 = C' \int_1^w \frac{du}{u \sqrt[n]{\prod_{r=1}^n \operatorname{sn}^2((K/\pi i)\ln(u/a_r); k)}}$$

or

$$z - 1 = C' \int_1^w \frac{du}{u \sqrt[n]{\prod_{r=-n}^{n-1} \operatorname{sn}((K/\pi i)\ln u - 2rK/n; k)}}.$$

The fact that

$$\operatorname{sn}(v + \alpha) \operatorname{sn}(v - \alpha) = \frac{\operatorname{sn}^2 v - \operatorname{sn}^2 \alpha}{1 - k^2 \operatorname{sn}^2 \alpha \operatorname{sn}^2 v},$$

allows us to write

$$\prod_{r=-n}^{n-1} \operatorname{sn}\left(v - \frac{2rK}{n}\right)$$

$$= - \operatorname{sn}^2 v \prod_{r=1}^{n-1} \frac{\operatorname{sn}^2 v - \operatorname{sn}^2(2rK/n)}{1 - k^2 \operatorname{sn}^2(2rK/n) \operatorname{sn}^2 v}.$$

But since

$$\operatorname{sn}^2 \frac{2\alpha K}{n} = \operatorname{sn}^2 \frac{2(n - \alpha)K}{n},$$

the quantity

$$\prod_{r=-n}^{n-1} \operatorname{sn}\left(v - \frac{2rK}{n}\right) = \Omega$$

is equal to

$$- \operatorname{sn}^2 v \prod_{r=1}^{(n-1)/2} \left\{ \frac{\operatorname{sn}^2 v - \operatorname{sn}^2(2rK/n)}{1 - k^2 \operatorname{sn}^2(2rK/n) \operatorname{sn}^2 v} \right\}^2$$

for odd n and to

$$\frac{\operatorname{sn}^2 v \operatorname{cn}^2 v}{\operatorname{dn}^2 v} \prod_{r=1}^{n/2-1} \left\{ \frac{\operatorname{sn}^2 v - \operatorname{sn}^2(2rK/n)}{1 - k^2 \operatorname{sn}^2(2rK/n) \operatorname{sn}^2 v} \right\}^2$$

$(^{25})$The constant C' has a different value in each formula.

for even n.

On the basis of Table XXII we conclude that in both cases

$$\Omega = N \operatorname{sn}^2(v/M; \lambda), \qquad L = K/nM, \qquad L' = K'/M,$$

where N is a constant, and λ and M can be determined with the help of the indicated formulas in Table XXII.

Thus the mapping function can be given the form

$$z - 1 = C \int_1^w \frac{du}{u \sqrt[n]{\operatorname{sn}^2((K \ln u)/M \pi i; \lambda)}},$$

where

$$\lambda = k^n \prod_{r=1}^{[n/2]} \operatorname{sn}^4\left(\frac{2r-1}{n} K; k\right),$$

$$M = \prod_{r=1}^{[n/2]} \frac{\operatorname{sn}^2((2r-1)K/n; k)}{\operatorname{sn}^2(2rK/n; k)},$$

and C is a constant.

It is useful to compare the result we have obtained with a function mapping the unit disk in the w-plane onto a regular n-gon inscribed in the unit disk in the z-plane. If the points $w = 0$ and $w = 1$ are to pass into the points $z = 0$ and $z = 1$, respectively, then for the mapping function we have that

$$z - 1 = C \int_1^w \frac{du}{\sqrt[n]{(u^n - 1)^2}}. \tag{2}$$

Observing that

$$u^{n/2} - u^{-n/2} = 2i \sin \frac{n \ln u}{2i},$$

we can rewrite (2) in the form

$$z - 1 = C \int_1^w \frac{du}{u \sqrt[n]{\sin^2((n \ln u)/2i)}}.$$

We see that in our case the function sn is where the ordinary sine is in this formula.

EXAMPLE 2. Map the plane, cut along two parallel segments symmetric with respect to the real axis, onto a circular annulus (Figure 22).

Suppose that the given segments $A_1 A_2$ and $B_1 B_2$ have length 2β and are at a distance α from the axis OX; we shall construct a function mapping the annulus

$$h^{1/2} \le |u| \le h^{-1/2} \qquad (0 < h < 1)$$

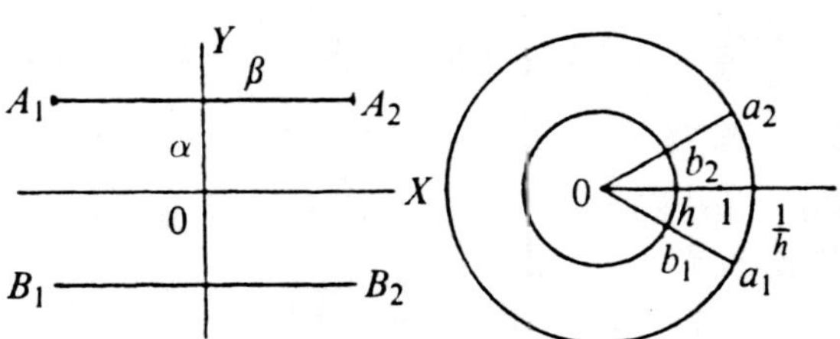

FIGURE 22

onto the plane, cut along A_1A_2 and B_1B_2. By symmetry, it can be assumed that the points

$$a_1 = h^{-1/2}e^{-i\mu}, \quad a_2 = 1/ha_1 \quad b_1 = ha_1, \quad b_2 = 1/a_1$$

are the inverse images of the endpoints A_1, A_2, B_1 and B_2 of the segments. Here the point $u = 1$ is the inverse image of the point at infinity in the z-plane.

Assuming the foregoing and setting $w = h^{1/2}u$ in the general formula in §48, we have that[26]

$$A'z + B' = \int \frac{\vartheta_1\left(\frac{\ln u - \ln a_1}{2\pi i}\right)\vartheta_1\left(\frac{\ln u + \ln a_1}{2\pi i}\right)}{\vartheta_1^2\left(\frac{\ln u}{2\pi i}\right)\vartheta_1^2\left(\frac{\ln u + \ln h}{2\pi i}\right)} \times \vartheta_1\left(\frac{\ln u - \ln ha_1}{2\pi i}\right)\vartheta_1\left(\frac{\ln u + \ln ha_1}{2\pi i}\right)\frac{du}{u^2},$$

where $\vartheta_1(v) = \vartheta_1(v|\tau)$ and $h = e^{\pi i\tau}$; or, by the formulas for reduction of theta functions,

$$A'z + B' = \int \frac{\vartheta_1\left(\frac{\ln u - \ln a_1}{2\pi i}\right)\vartheta_0\left(\frac{\ln u - \ln a_1}{2\pi i}\right)}{\vartheta_1^2\left(\frac{\ln u}{2\pi i}\right)\vartheta_0^2\left(\frac{\ln u}{2\pi i}\right)} \times \vartheta_1\left(\frac{\ln u + \ln a_1}{2\pi i}\right)\vartheta_0\left(\frac{\ln u + \ln a_1}{2\pi i}\right)\frac{du}{u}.$$

Further, if we use the identity

$$\vartheta_1(v|\tau)\vartheta_0(v|\tau) = \tfrac{1}{2}\vartheta_2(0|\tau/2)\vartheta_1(v|\tau/2)$$

(see Table XXI), we get

$$A'z + B'$$
$$= \int \frac{\vartheta_1\left(\frac{\ln u + \ln a_1}{2\pi i}\Big|\frac{\tau}{2}\right)\vartheta_1\left(\frac{\ln u - \ln a_1}{2\pi i}\Big|\frac{\tau}{2}\right)}{\vartheta_1^2\left(\frac{\ln u}{2\pi i}\Big|\frac{\tau}{2}\right)}\frac{du}{u}. \tag{3}$$

---

[26] The constants A' and B' have different values in the different formulas.

Consider the expression

$$\frac{\vartheta_1(v + c|\tau/2)\vartheta_1(v - c|\tau/2)}{\vartheta_1^2(v|\tau/2)}.$$

This is an elliptic function with periods 1 and $\tau/2$. It is easy to see that the decomposition of this function into partial fractions can be written in the form

$$\frac{\vartheta_1(v + c)\vartheta_1(v - c)}{\vartheta_1^2(v)} = L\left[\frac{\vartheta_1'(v)}{\vartheta_1(v)}\right]' + M\frac{\vartheta_1'(v)}{\vartheta_1(v)} + N,$$

where L, M, and N are constants. Since the second term on the right-hand side is an odd function of v and all the remaining terms are even, it follows that $M = 0$. Expanding both sides in powers of v^2, we get that

$$-\frac{\vartheta_1^2(c)}{\vartheta_1'^2(0)}\frac{1}{v^2} + \frac{\vartheta_1^2(c)\vartheta_1'''(0)}{3\vartheta_1'^3(0)} + \frac{\vartheta_1'^2(c) - \vartheta_1(c)\vartheta_1''(c)}{\vartheta_1'^2(0)} + \cdots$$

$$= L\left\{-\frac{1}{v^2} + \frac{\vartheta_1'''(0)}{3\vartheta_1'(0)} + \cdots\right\} + N.$$

From this,

$$L = \frac{\vartheta_1^2(c)}{\vartheta_1'^2(0)}, \qquad N = \frac{\vartheta_1'^2(c) - \vartheta_1(c)\vartheta_1''(c)}{\vartheta_1'^2(0)}.$$

Applying this decomposition into partial fractions to the integral (3), we get the following representation of the mapping function:

$$A'z + B' = 2\pi i\frac{\vartheta_1^2(\frac{\ln a_1}{2\pi i}|\frac{\tau}{2})\vartheta_1'(\frac{\ln u}{2\pi i}|\frac{\tau}{2})}{\vartheta_1'^2(0|\frac{\tau}{2})\vartheta_1(\frac{\ln u}{2\pi i}|\frac{\tau}{2})}$$

$$+ \frac{\vartheta_1'^2(\frac{\ln a_1}{2\pi i}|\frac{\tau}{2}) - \vartheta_1(\frac{\ln a_1}{2\pi i}|\frac{\tau}{2})\vartheta_1''(\frac{\ln a_1}{2\pi i}|\frac{\tau}{2})}{\vartheta_1'^2(0|\frac{\tau}{2})}\ln u.$$

Since the mapping function is single-valued interior to the annulus, the second term on the right-hand side must be equal to zero. This gives us that

$$Az + B = \vartheta_1'\left(\frac{\ln u}{2\pi i}\bigg|\frac{\tau}{2}\right)\bigg/\vartheta_1\left(\frac{\ln u}{2\pi i}\bigg|\frac{\tau}{2}\right),$$

and

$$\vartheta_1'^2\left(\frac{\ln a_1}{2\pi i}\bigg|\frac{\tau}{2}\right) - \vartheta_1\left(\frac{\ln a_1}{2\pi i}\bigg|\frac{\tau}{2}\right)\vartheta_1''\left(\frac{\ln a_1}{2\pi i}\bigg|\frac{\tau}{2}\right) = 0. \tag{4}$$

It is clear from symmetry considerations that the point $z = 0$ passes into the point $u = -1$. Therefore,

$$B = \vartheta_1'\left(\frac{1}{2}\bigg|\frac{\tau}{2}\right)\bigg/\vartheta_1\left(\frac{1}{2}\bigg|\frac{\tau}{2}\right) = 0.$$

On the other hand, the point $u = a_2 = h^{-1/2}e^{i\mu}$ must correspond to the point $z = \beta + i\alpha$. Consequently,

$$
\begin{aligned}
A(\beta + i\alpha) &= -\vartheta_1'\left(\frac{\tau}{4} - \frac{\mu}{2\pi}\bigg|\frac{\tau}{2}\right)\bigg/\vartheta_1\left(\frac{\tau}{4} - \frac{\mu}{2\pi}\bigg|\frac{\tau}{2}\right) \\
&= \pi i + \vartheta_0'\left(\frac{\mu}{2\pi}\bigg|\frac{\tau}{2}\right)\bigg/\vartheta_0\left(\frac{\mu}{2\pi}\bigg|\frac{\tau}{2}\right),
\end{aligned}
$$

and, similarly,

$$
A(-\beta + i\alpha) = \pi i - \vartheta_0'\left(\frac{\mu}{2\pi}\bigg|\frac{\tau}{2}\right)\bigg/\vartheta_0\left(\frac{\mu}{2\pi}\bigg|\frac{\tau}{2}\right).
$$

Hence $A = \pi/\alpha$ and

$$
\pi\frac{\beta}{\alpha} = \vartheta_0'\left(\frac{\mu}{2\pi}\bigg|\frac{\tau}{2}\right)\bigg/\vartheta_0\left(\frac{\mu}{2\pi}\bigg|\frac{\tau}{2}\right). \tag{5}
$$

Thus, the mapping function has the form

$$
z = \frac{\alpha}{\pi}\vartheta_1'\left(\frac{\ln u}{2\pi i}\bigg|\frac{\tau}{2}\right)\bigg/\vartheta_1\left(\frac{\ln u}{2\pi i}\bigg|\frac{\tau}{2}\right),
$$

and it remains to write the two equations serving for the determination of μ and τ. Equality (5) is one. The second is obtained from (4) and can be written in the form

$$
\pi\frac{\beta}{\alpha} = \vartheta_0''\left(\frac{\mu}{2\pi}\bigg|\frac{\tau}{2}\right)\bigg/\vartheta_0'\left(\frac{\mu}{2\pi}\bigg|\frac{\tau}{2}\right). \tag{6}
$$

Equations (5) and (6) can be represented in another form more convenient for computations. We present the result without proof. It consists in the following: let

$$
q = e^{\pi i \tau/2}, \quad \sqrt{k} = \frac{2(q^{1/4} + q^{9/4} + \cdots)}{1 + 2q + \cdots}, \quad \lambda = \frac{1}{k}\sqrt{\frac{K-E}{K}};
$$

then

$$
\mu = \frac{\pi}{K}\int_0^\lambda \frac{dt}{\sqrt{(1 - t^2)(1 - k^2 t^2)}}
$$

and

$$
\beta = \frac{2\alpha}{\pi}\left\{K\int_0^\lambda \sqrt{\frac{1 - k^2 t^2}{1 - t^2}}\,dt - E\int_0^\lambda \frac{dt}{\sqrt{(1 - t^2)(1 - k^2 t^2)}}\right\}.
$$

Thus, it is possible to assign the quantity q, i.e., the radii of the boundaries of the annulus. From the chosen q we find both μ and the ratio of β to α with the help of tables. This enables us to construct (for example, graphically) the dependence between q and β/α. Having this dependence, we can take the ratio β/α as the original parameter.

EXAMPLE 3. Map the z-plane, cut along two given segments of the real axis, onto a circular annulus in the w-plane.

Suppose that these segments of the real axis in the z-plane are $[-1, \alpha]$ and $[\beta, 1]$, where $-1 < \alpha < \beta < 1$. Let

$$k^2 = \frac{2(\beta - \alpha)}{(1 - \alpha)(1 + \beta)} \tag{7}$$

and take k $(0 < k < 1)$ as the modulus of the elliptic functions.

We next determine the number ρ from

$$1 - 2\,\mathrm{sn}^2\rho = \alpha \tag{8}$$

under the addition condition $0 < \rho < K$.

It follows from (7) and (8) that

$$2cn^2\rho/dn^2\rho - 1 = \beta.$$

Now let

$$z = \frac{\mathrm{sn}^2 u\,\mathrm{cn}^2\rho + \mathrm{cn}^2 u\,\mathrm{sn}^2\rho}{\mathrm{sn}^2 u - \mathrm{sn}^2\rho},$$

which can be rewritten in the form

$$z - \alpha = \frac{1 - \alpha^2}{2\,\mathrm{sn}^2 u + \alpha - 1}. \tag{9}$$

Next, take the rectangle Δ in the u-plane defined by the inequalities

$$-K \le \Re u \le 0, \qquad -K' \le \Im u \le K'.$$

The real axis divides this rectangle into two rectangles: an upper and a lower. Let us traverse the boundary of the upper rectangle in the positive direction, starting from the point $u = -\rho$. It is easy to see that the point z then describes the real axis from $-\infty$ to ∞. With the help of (9) this upper rectangle can therefore be mapped onto the upper half of the z-plane. Here the part of the real axis in the z-plane obtained by removing the segment $[-1, 1]$ corresponds to the lower base of the rectangle. Further, by the same formula (9), the lower rectangle is mapped onto the lower half of the z-plane.

Therefore, our formula (9) maps the whole rectangle Δ onto the whole z-plane, cut along $[-1, 1]$. Now let $u = K'\pi^{-1}\ln w$ under the additional condition that $u = 0$ for $w = 1$. With the help of this formula the rectangle Δ is mapped onto the annulus in the w-plane bounded by the circles $|w| = 1$ and $|w| = e^{-\pi K/K'}$ and cut along a segment of the negative half of the real axis. Removal of this cut is equivalent in the u-plane to identification of the upper side of Δ with the lower side, and in the z-plane to sewing

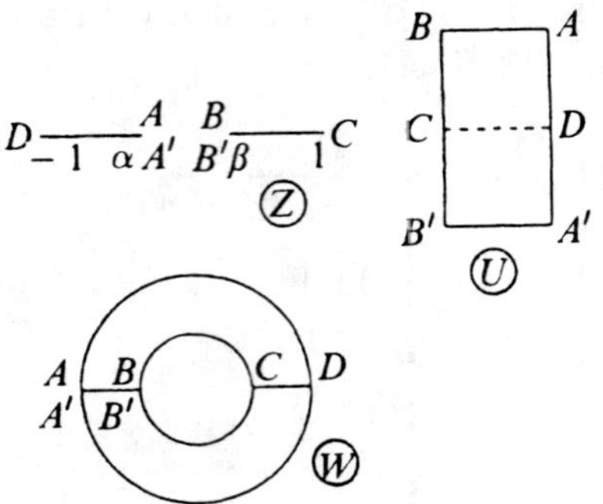

FIGURE 23

together the upper half-plane and the lower half-plane along the segment
$[\alpha, \beta]$ of the real axis (Figure 23).

This implies that the formula

$$z = \alpha + \frac{1 - \alpha^2}{2\,\mathrm{sn}^2((K' \ln w)/\pi) + \alpha - 1}$$

gives the desired mapping.

CHAPTER 9

Extremal Properties of Fractions to Which a Transformation of Elliptic Functions Reduces

§50. Statement of the problems

In 1877 the great article of E. I. Zolotarev, *Application of elliptic functions to questions of functions deviating least and most from zero*, appeared in [20].

At the beginning of his article Zolotarev writes:

"In spite of the presently known applications (which are remarkable to the highest degree) of elliptic functions to number theory, geometry, and mechanics, I submit that the theory of elliptic functions still leaves much to be desired from the aspect of applications.

"Therefore, I did not regard it as superfluous to consider certain smallest-quantity problems that can be solved with the help of basic formulas in the theory of elliptic functions. These questions belong to the class of smallest-quantity problems for which methods of solution were first given by P. L. Tchebycheff."

Zolotarev poses and solves four problems. The first two are about polynomials, while the third and fourth are about rational fractions. The last two problems, which are especially interesting in a mathematical respect, are also of great significance for certain computations in modern electrical engineering[27]. These problems are closely related to a number of other problems, and, in particular, to a certain problem of Tchebycheff, which he solved twelve years after Zolotarev.

In this section we formulate the problems and also establish connections between them[28]. The next section will be devoted to the solutions.

The deviation of a continuous function $g(x)$ from a continuous function $f(x)$ on a finite or infinite closed point set $\mathscr{E}$ on the number line is defined

---

[27] We have in mind the work of Cauer on the theory and computation of electrical filters. See Cauer [17], and also Taft [18].

[28] Here I use my paper [22] and my two notes [24] and [25].

141

to be the quantity

$$\sup_{g} |f(x) - g(x)|.$$

For our purposes it is convenient to regard the union of two intervals $[-\infty, \alpha]$ and $[\beta, \infty]$ with $\alpha < \beta$ as a single (improper) interval. We agree to denote this interval by $[\beta, \alpha]$. Thus, $[a, b]$ is an ordinary interval if $a < b$ and an improper interval if $a > b$.

Let I_1 and I_2 be disjoint closed intervals on the real axis. One of them may contain the point at infinity in its interior or on its boundary.

PROBLEM A*. Among all real functions $\varphi(x)/\psi(x)$ with $\varphi(x)$ and $\psi(x)$ polynomials of degree n find those that deviate least from the function

$$h(x) = \begin{cases} -1 & (x \in I_1), \\ 1 & (x \in I_2) \end{cases}$$

on the point set consisting of the intervals I_1 and I_2.

PROBLEM B*. A closed finite or infinite interval E of the number line not containing 0 is given. The collection of all rational fractions of degree n taking values in E on the interval I_2 is considered. Among these fractions it is required to find those deviating least from zero on I_1.

In essence, Problem A* coincides with Zolotarev's fourth problem, and Problem B* with the third.

PROBLEM C*. Among all real functions $\Phi(x)/\Psi(x)$ with $\Phi(x)$ and $\Psi(x)$ polynomials of degree r find those for which the logarithm of the ratio

$$\sqrt{x} : \frac{\Phi(x)}{\Psi(x)}$$

deviates least from 0 in a given interval $[1, 1/k^2]$, where $0 < k < 1$.

This is the problem of Tchebycheff.

Arbitrary intervals are given in Problems A* and B*. However, by subjecting x to a linear fractional transformation we can introduce some completely definite intervals in place of these intervals.

Therefore, instead of Problem A* we can take the following problem.

PROBLEM A. Among all real functions $y = \varphi(x)/\psi(x)$ with $\varphi(x)$ and $\psi(x)$ polynomials of degree n find those deviating least from the function

$$\operatorname{sgn} x = \begin{cases} -1 & (x < 0) \\ 1 & (x > 0) \end{cases}$$

on the point set consisting of the two intervals

$$[-1/k, -1], \quad [1, 1/k] \quad (0 < k < 1) \tag{1}$$

Problem B* specifies an inequality that must be satisfied by a function on the interval I_2. By also subjecting the function to a linear fractional transformation we can reduce Problem B* to the following problem.

PROBLEM B. Among all real rational fractions $z = f(t)/g(t)$ of nth degree satisfying $|z| \geq 1$ on the interval $[1/\kappa, -1/\kappa]$ $(0 < x < 1)$, find the one that deviates least from zero on $[-1, 1]$.

We show that one of Problems A and B can be reduced to the other.

Suppose that the irreducible fraction $z = f_0(t)/g_0(t)$ is a solution of Problem B. It is easy to see that the polynomial $f_0(t)$ is precisely of degree n. Indeed, if the degree of $f_0(t)$ were less than n, then the function $\tilde{z} = \kappa t f_0(t)/g_0(t)$ would also be a rational fraction of degree n, and we also would have the inequality $|\tilde{z}| \geq 1$ on $[1/\kappa, -1/\kappa]$.

At the same time,

$$\max |\tilde{z}| \leq \kappa \max |z| < \max |z|$$

in $[-1, 1]$, and hence z could not be a solution of Problem B.

It is also easy to see that

$$\min_{[1/\kappa, -1/\kappa]} |z| = 1.$$

Let

$$\max_{[-1,1]} |z| = m.$$

The number m is certainly less than 1, as shown by the function $z = \kappa t$, which satisfies the conditions of Problem B. Let

$$y = \frac{1-m}{1+m} \frac{z - \sqrt{m}}{z + \sqrt{m}},$$

$$x = \frac{1+\sqrt{\kappa}}{1-\sqrt{\kappa}} \frac{t\sqrt{\kappa} - 1}{t\sqrt{\kappa} + 1},$$

$$k = \left(\frac{1-\sqrt{\kappa}}{1+\sqrt{\kappa}}\right)^2.$$

If $-1 \leq t \leq 1$, then $-1/k \leq x \leq -1$; similarly, if $|t| \geq 1/\kappa$, then $1 \leq x \leq 1/k$. Thus, we have two intervals on the x-axis:

$$[-1/k, -1], \qquad [1, 1/k]. \tag{1}$$

In the first interval $\max |z| = m$ by assumption. But since

$$y + 1 = \frac{2(z + m\sqrt{m})}{(1+m)(z + \sqrt{m})},$$

it follows that

$$\max |y + 1| = 2\sqrt{m}/(1 + m)$$

in the first interval. As we know, $\min |z| = 1$ in the second interval. Therefore,

$$\max |y - 1| = 2\sqrt{m}/(1 + m)$$

in the second interval. We see that the deviation of the function y from $\operatorname{sgn} x$ in the intervals (1) is equal to

$$\mu = 2\sqrt{m}/(1 + m);$$

μ is a monotonically increasing function of m in $(0, 1)$.

From this we see that $y = y(x)$ is a solution of Problem A. Thus, having a solution of Problem B, we easily get a solution of Problem A. But, conversely, if a solution $y = y(x)$ of Problem A is known, then it is easy to find a solution $z = z(t)$ of Problem B.

It follows from the above conclusion about the degree of $f_0(t)$ being precisely n that if an irreducible fraction $y = \varphi(x)/\psi(x)$ is a solution of Problem A, then at least one of the functions $\varphi(x)$ and $\psi(x)$ is precisely of degree n.

Suppose that the irreducible fraction $y = \varphi(x)/\psi(x)$ is a solution of Problem A. Denote by

$$x_1 < x_2 < \cdots < x_q \tag{2}$$

the successive points belonging to the intervals (1) at which the difference $y - \operatorname{sgn} x$ takes its maximal numerical value μ with alternating signs in these intervals:

$$\mu = \max |y - \operatorname{sgn} x|$$

(this number μ is obviously less than 1).

We show that the number q cannot be less than $2n + 2$.

Suppose that x_p in the series (2) is the last point in the closed interval $[-1/k, -1]$, while x_{p+1} is the first point in $[1, 1/k]$. In this case the quantities $y(x_p) + 1$ and $y(x_{p+1}) - 1$ have opposite signs.

With the help of the points $\xi_1 < \cdots < \xi_{p-1}$ and $\xi_{p+1} < \cdots < \xi_{q-1}$ we break up the respective intervals $[-1/k, -1]$ and $[1, 1/k]$ into subintervals in which, successively, one of the inequalities

$$-\mu \le y - \operatorname{sgn} x < \mu - \alpha,$$

$$-\mu + \alpha < y - \operatorname{sgn} x \le \mu$$

holds, where α is a positive number. Then we construct the function

$$\Omega(x) = (x - \xi_1)(x - \xi_2) \cdots (x - \xi_{p-1})x$$
$$\times (x - \xi_{p+1})(x - \xi_{p+2}) \cdots (x - \xi_{q-1}),$$

which has degree $q - 1 \le 2n$ if, contrary to what is to be proved, $q < 2n + 2$.

Since $\varphi(x)$ and $\psi(x)$ are relatively prime and at least one of them is precisely of degree n, there exist polynomials $\varphi_1(x)$ and $\psi_1(x)$ of degree at most n such that

$$\Omega(x) = \varphi_1(x)\psi(x) - \varphi(x)\psi_1(x).$$

We now introduce the function

$$\widetilde{y} = \frac{\varphi(x) - \theta\varphi_1(x)}{\psi(x) - \theta\psi_1(x)},$$

where θ is a real parameter, and we consider the expression

$$\widetilde{y} - \operatorname{sgn} x = \frac{\varphi(x) - \theta\varphi_1(x)}{\psi(x) - \theta\psi_1(x)} - \frac{\varphi(x)}{\psi(x)} + \{y - \operatorname{sgn} x\}$$

$$= \{y - \operatorname{sgn} x\} - \frac{\theta\Omega(x)}{\psi(x)[\psi(x) - \theta\psi_1(x)]}.$$

By an assumption of the problem $\psi(x)$ does not vanish in the intervals (1). Therefore, for sufficiently small $|\theta|$

$$\psi(x)[\psi(x) - \theta\psi_1(x)] > \sqrt{|\theta|}.$$

By virtue of the character of the function $\Omega(x)$ it is possible to choose the sign of θ with sufficiently small modulus in such a way that in the intervals (1)

$$|\widetilde{y} - \operatorname{sgn} x| < \mu,$$

and this proves our assertion.

We now dwell on some corollaries.

a) If $R(x) = \varphi(x)/\psi(x)$ is a solution of Problem A, then the equation

$$\{R(x) - \operatorname{sgn} x\}^2 = \mu^2 \tag{3}$$

has simple roots $-1/k$, -1, 1, and $1/k$; nevertheless, the roots of this equation lying interior to the intervals (1) (the total number of these roots is $2n - 2$) are double roots.

b) If $R(x) = \varphi(x)/\psi(x)$ is a solution of Problem A, then the interval $(-1, 1)$ can contain roots of only one of the equations $\varphi(x) = 0$ or $\psi(x) = 0$.

Indeed, otherwise the function $|R(x)|$ would take the values $1 - \mu$ and $1 + \mu$ in $(-1, 1)$, and then the number of roots of (3) in the intervals (1) would be less than $2n + 2$.

We note that $(1 - \mu^2)\psi(x)/\varphi(x)$ is a solution of Problem A if $\varphi(x)\psi(x)$ is.

In view of b), this implies that we can confine ourselves to the solutions that do not become infinite in $(-1, 1)$.

Such a finite solution is unique. Indeed, suppose that there are two solutions

$$\varphi_1(x)/\psi_1(x), \qquad \varphi_2(x)/\psi_2(x), \tag{4}$$

and let $x_1 < x_2 < \cdots < x_{n+2}$ be the "deviation points" for the first solutions. We take the difference

$$\begin{aligned}
\Delta(x) &= \frac{\varphi_1(x)}{\psi_1(x)} - \frac{\varphi_2(x)}{\psi_2(x)} \\
&= \left\{ \frac{\varphi_1(x)}{\psi_1(x)} - \operatorname{sgn} x \right\} - \left\{ \frac{\varphi_2(x)}{\psi_2(x)} - \operatorname{sgn} x \right\} \\
&= \Delta_1(x) - \Delta_2(x).
\end{aligned}$$

Let

$$\Delta_1(x_1) = \varepsilon\mu, \ \Delta_1(x_2) = -\varepsilon\mu, \ \Delta_1(x_3) = \varepsilon\mu, \ldots,$$

where $\varepsilon = \pm 1$. On the other hand,

$$|\Delta_2(x_k)| \leq \mu \qquad (k = 1, 2, \ldots, 2n + 2).$$

Therefore, $\Delta(x_k)$ either is equal to zero or has the same sign as $\Delta_1(x_k)$. On the basis of this it is easy to see that $\Delta(x)$ has at least $2n + 1$ roots in $[-1/k, 1/k]$. But since

$$\Delta(x) = \frac{\varphi_1(x)\psi_2(x) - \varphi_2(x)\psi_1(x)}{\psi_1(x)\psi_2(x)},$$

where the numerator $\varphi_1(x)\psi_2(x) - \varphi_2(x)\psi_1(x)$ is a polynomial of degree $2n$ it follows that $\Delta(x)$ is identically zero, i.e., the solutions in (4) are not different.

It is now easy to see that only an odd function can be a solution of Problem A. Indeed, let

$$R(x) = \varphi(x)/\psi(x) \tag{5}$$

be a solution of Problem A that remains finite in $(-1, 1)$. But then the function

$$-R(-x) = -\varphi(-x)/\psi(-x) \tag{6}$$

is also such a solution. Consequently, the functions (5) and (6) are identical.

The facts we have established enable us show without difficulty that Problems A and C* are equivalent.

With this goal we observe that if

$$\max |\ln Y| = \ln H,$$

then

$$\max \left| 1 - \frac{2H}{H^2 + 1} Y \right| = \frac{H^2 - 1}{H^2 + 1},$$

and if

$$\max |1 - y| = G < 1,$$

then

$$\max \left| \ln \frac{y}{\sqrt{1 - G^2}} \right| = \frac{1}{2} \ln \frac{1 + G}{1 - G}.$$

Therefore, the following problem is equivalent to Problem C*.

PROBLEM C. Among all real functions

$$Y = \sqrt{x}\, \Psi(x)/\Phi(x),$$

with $\Phi(x)$ and $\Psi(x)$ polynomials of degree r find the one that deviates least from 1 on $[1, 1/k^2]$.

Now let $x = X^2$. Then instead of $[1, 1/k^2]$ we can take the two intervals $[-1/k, -1]$ and $[1, 1/k]$. Our function becomes

$$Y = X\Psi(X^2)/\Phi(X^2),$$

and the quantity

$$\max_{[1,1/k^2]} \left| 1 - \frac{\sqrt{x}\, \Psi(x)}{\Phi(x)} \right|$$

is clearly equal to

$$\max_{[-1/k,-1],[1,1/k]} \left| \operatorname{sgn} X - \frac{X\Psi(X^2)}{\Phi(X^2)} \right|.$$

We have arrived at Problem A under the condition that $n = 2r$ or $n = 2r + 1$.

The converse (from Problem A to Problem C) is even simpler. By the considerations in this section, everything has been reduced to the solution of Problem C.

§51. Solution of Problem C

The solution of Problem C can be obtained with the help of a general theorem of Tchebycheff consisting in the following.

Let $[a, b]$ be a finite closed interval, and let $f(x)$ and $s(x)$ be two continuous functions on it, the second nonvanishing. We consider expressions of the form

$$W(x) = s(x)\frac{q_0 x^n + q_1 x^{n-1} + \cdots + q_n}{p_0 x^m + p_1 x^{m-1} + \cdots + p_m},$$

where m and n are given. Among these functions $W(x)$ there exists one deviating least from $f(x)$ on $[a, b]$, and it is unique if we do not distinguish between two fractions that coincide after reduction.

If this function has the form

$$P(x) = s(x)\frac{B(x)}{A(x)}$$

$$= s(x)\frac{b_0 x^{n-\nu} + b_1 x^{n-\nu-1} + \cdots + b_{n-\nu}}{a_0 x^{m-\mu} + a_1 x^{m-\mu-1} + \cdots + a_{m-\mu}},$$

where $0 \le \mu \le m$, $0 \le \nu \le n$, $a_0 \ne 0$, and the fraction $B(x)/A(x)$ is irreducible, then the number of successive points of $[a, b]$ at which $f(x) - P(x)$ takes its maximal numerical value on $[a, b]$ with alternating signs is not less than $m + n - d + 2$, where $d = \min\{\mu, \nu\}$. This property completely characterizes the extremal function $P(x)$.

Only the part of the theorem asserting that $P(x)$ deviates least from $f(x)$ under the indicated characteristic condition is essential for what follows. This can easily be proved by the same method we used in §50 to prove for the special case of Problem A that it has at most one solution.

We need not prove the existence of a solution, since we actually construct this solution for the case we are interested in.[29]

In Problem C the interval $[a, b]$ is $[1, 1/k^2]$, and also $f(x) = 1$, $s(x) = \sqrt{x}$, and $m = n = r$. We present a solution of the problem in parametric form. It is given by the formulas

$$x = \operatorname{sn}^2(u; k), \qquad Y = \frac{2\lambda}{1+\lambda} \operatorname{sn}\left(\frac{u}{M}; \lambda\right),$$

where $L = K/M$ and $L' = K'/(2r+1)M$.

Thus, Y is the function of x obtained by dividing the second period by $2r + 1$.

Proceeding to the proof, we turn first to Table XXIII, according to which

$$Y = \frac{2\lambda}{1+\lambda}\frac{\operatorname{sn}(u; k)}{M} \prod_{\alpha=1}^{r} \frac{1 + \operatorname{sn}^2(u; k)/c_{2\alpha}}{1 + \operatorname{sn}^2(u; k)/c_{2\alpha-1}},$$

where

$$c_\alpha = \operatorname{sn}^2\left(\frac{\alpha}{2r+1}K'; k'\right) \Big/ \operatorname{cn}^2\left(\frac{\alpha}{2r+1}K'; k'\right).$$

Therefore,

$$Y = \frac{2\lambda}{1+\lambda}\frac{\sqrt{x}}{M} \prod_{\alpha=1}^{r} \frac{1 + x/c_{2\alpha}}{1 + x/c_{2\alpha-1}},$$

which implies that Y has the required form.

We now consider the difference $1 - Y$ as x runs through $[1, 1/k^2]$.

---

[29] The reader can find details about this theorem in my book [19].

Let $u = K + iv$; then

$$x = \operatorname{sn}^2(u; k) = \frac{\operatorname{cn}^2(iv; k)}{\operatorname{dn}^2(iv; k)} = \frac{1}{\operatorname{dn}^2(v; k')}.$$

We let v increase from 0 to K'. Then $\operatorname{dn}^2(v; k')$ decreases from 1 to $1 - k'^2 = k^2$, and hence x increases from 1 to $1/k^2$. This is the interval in which x varies.

Consider the difference $\Delta(x) = 1 - Y$. It is equal to

$$\Delta(x) = 1 - \frac{2\lambda}{1 + \lambda} \operatorname{sn}\left(\frac{K + iv}{M}; \lambda\right)$$

$$= 1 - \frac{2\lambda}{1 + \lambda} \frac{1}{\operatorname{dn}(v/M; \lambda')} = 1 - \frac{2\lambda}{1 + \lambda} \frac{1}{\operatorname{dn}(w; \lambda')}.$$

Suppose that v increases from 0 to K'. This means that w increases from 0 to $(2r+1)L'$. The function $\operatorname{dn}(w; \lambda')$ is always between 1 and $\sqrt{1 - \lambda'^2} = \lambda$. Further, it has values $1, \lambda, 1, \ldots, 1, \lambda$ at the points $w = 0, L', 2L', \ldots, 2rL'$, $(2r + 1)L'$. The corresponding values of $\Delta(x)$ are equal to

$$\frac{1 - \lambda}{1 + \lambda}, -\frac{1 - \lambda}{1 + \lambda}, \ldots, -\frac{1 - \lambda}{1 + \lambda},$$

Thus, the difference $1 - Y$ takes its maximal numerical value $(1 - \lambda)/(1 + \lambda)$ at the $2r + 2$ successive points

$$x_0 = 1, \; x_1 = \frac{1}{\operatorname{dn}^2(\frac{K'}{2r+1}; k')}, \; x_2 = \frac{1}{\operatorname{dn}^2(\frac{2K'}{2r+1}; k')}, \ldots$$

of the interval $[1, 1/k^2]$.

The proof is complete.

A number of other problems can be solved by starting from the solution of Problem C and using suitable linear fractional transformations. We confine ourselves here to a summary of some results.

Let us consider real rational functions $R_{i,j}(x)$ with numerator of degree i and denominator of degree j. Among all these functions it is required to find those for which $s(x)R_{i,j}(x)$ deviates least from 1 in $[a, b]$. Thus, we are concerned with an approximation of $1/s(x)$ by the function $R_{i,j}(x)$ such that them maximum on $[a, b]$ of the relative error has minimal value.

The given problems are contained in Table 1.

The desired solution is denoted by y; moreover, let

$$G = \max_{[a,b]} |1 - s(x)y|.$$

The solutions of these problems are contained in Table 2. Here λ (respectively, λ_1) denotes the modulus of the elliptic functions at which we arrive when we divide the first (respectively, second) period by n.

TABLE 1.

No.	$s(x)$	a	b	i	j
1	$\sqrt{1 - k^2 x}$	0	1	m	m
2	$\sqrt{1 - k^2 x}$	0	1	$m - 1$	m
3	$\sqrt{\frac{x}{x-1}}$	$\frac{1}{k^2}$	∞	m	m
4	$\sqrt{x(x - 1)}$	$\frac{1}{k^2}$	∞	$m - 1$	m
5	$\sqrt{\frac{x+1}{x-1}}$	$\frac{1}{k}$	$-\frac{1}{k}$	m	m
6	$\sqrt{\frac{1+kx}{1-kx}}$	-1	1	m	m
7	$\sqrt{x}$	1	$\frac{1}{k^2}$	m	m
8	$\sqrt{x}$	1	$\frac{1}{k^2}$	$m - 1$	m

TABLE 2.

No.	y	n	G
1	$\frac{2}{1+\lambda'} \prod_{\alpha=1}^{m} \frac{1-k^2 \operatorname{sn}^2(\frac{2\alpha-1}{n}K;k)x}{1-k^2 \operatorname{sn}^2(\frac{2\alpha}{n}K;k)x}$	$2m + 1$	$\frac{1-\lambda'}{1+\lambda'}$
2	$\frac{2\lambda'}{1+\lambda'} \frac{\prod_{\alpha=1}^{m-1} 1-k^2 \operatorname{sn}^2(\frac{2\alpha}{n}K;k)x}{\prod_{\alpha=1}^{m} 1-k^2 \operatorname{sn}^2(\frac{2\alpha-1}{n}K;k)x}$	$2m$	$\frac{1-\lambda'}{1+\lambda'}$
3	$\frac{2\lambda'}{1+\lambda'} \prod_{\alpha=1}^{m} \frac{\operatorname{sn}^2(\frac{2\alpha}{n}K;k)-x}{\operatorname{sn}^2(\frac{2\alpha-1}{n}K;k)-x}$	$2m + 1$	$\frac{1-\lambda'}{1+\lambda'}$
4	$-\frac{2\lambda'}{1+\lambda'} \frac{\prod_{\alpha=1}^{m-1} \operatorname{sn}^2(\frac{2\alpha}{n}K;k)-x}{\prod_{\alpha=1}^{m} \operatorname{sn}^2(\frac{2\alpha-1}{n}K;k)-x}$	$2m$	$\frac{1-\lambda'}{1+\lambda'}$
5	$\lambda' \prod_{\alpha=1}^{m} \frac{x+\operatorname{sn}(K-\frac{4\alpha}{n}K;k)}{x-\operatorname{sn}(K-\frac{4\alpha}{n}K;k)}$	$2m + 1$	λ
6	$\lambda' \prod_{\alpha=1}^{m} \frac{1+kx\,\operatorname{sn}(K-\frac{4\alpha}{n}K;k)}{1-kx\,\operatorname{sn}(K-\frac{4\alpha}{n}K;k)}$	$2m + 1$	λ
7[30]	$\frac{2\lambda_1}{1+\lambda_1} \frac{1}{M_1} \prod_{\alpha=1}^{m} \frac{1+\frac{x}{c_{2\alpha}}}{1+\frac{x}{c_{2\alpha-1}}}$	$2m + 1$	$\frac{1-\lambda_1'}{1+\lambda_1'}$
8[30]	$\frac{2\lambda_1}{1+\lambda_1} \frac{1}{M_1} \frac{\prod_{\alpha=1}^{m-1} 1+\frac{x}{c_{2\alpha}}}{\prod_{\alpha=1}^{m} 1+\frac{x}{c_{2\alpha-1}}}$	$2m$	$\frac{1-\lambda_1'}{1+\lambda_1'}$

---

[30] Here

$$c_\alpha = \frac{\operatorname{sn}^2(\alpha K'/n; k')}{\operatorname{cn}^2(\alpha K'/n; k')}$$

and

$$M_1 = \prod_{\alpha=1}^{m} \frac{\operatorname{sn}^2((2\alpha - 1)K'/n; k')}{(2\alpha/nK'; k')}.$$

CHAPTER 10

Generalization of Tchebycheff Polynomials

§52. Polynomials deviating least from zero

The Tchebycheff polynomials $T_n(x)$ $(n = 0, 1, 2, \ldots)$ are defined as follows:

$$T_n(x) = \cos n\varphi, \qquad x = \cos\varphi \quad (n = 0, 1, 2, \ldots).$$

Thus, $T_0(x) = 1$, $T_1(x) = x$, and

$$T_{n+1}(x) + T_{n-1}(x) = 2x T_n(x) \qquad (n = 1, 2, 3, \ldots),$$

from which it is clear that

$$T_n(x) = 2^{n-1} x^n + \cdots \qquad (n = 1, 2, 3, \ldots).$$

Let

$$L_n = \begin{cases} 1 & (n = 0), \\ 2^{1-n} & (n = 1, 2, 3, \ldots). \end{cases}$$

The following extremal property completely characterizes Tchebycheff polynomials: $L_n T_n(x)$ *deviates least from zero on* $[-1, 1]$ *among all polynomials of degree n with leading coefficient* 1 (consequently, the least deviation is equal to L_n).

This fact, which follows from the general Tchebycheff theorem (see §51), is easy to prove directly. Indeed, suppose that on $[-1, 1]$ the polynomial $P_n(x) = x^n + \cdots$ has deviation $\leq L_n$ from zero. Then at the points $x_k = \cos((n - k)\pi/n)$ $(k = 0, 1, \ldots, n)$ the difference $L_n T_n(x) - P_n(x) = Q(x)$, which is a polynomial of degree $\leq n - 1$, takes the values

$$Q(x_k) = (-1)^{n-k} L_n - P_n(x_k) = (-1)^{n-k} [L_n \pm |P_n(x_k)|],$$

where the expression in the square brackets is ≥ 0. From this it is easy to see that $Q(x)$ vanishes at least n times in $[-1, 1]$, and hence $Q(x) \equiv 0$.

The polynomials $T_n(x)$ have found diverse applications in engineering computations.[31]

---

[31]In 1952 the American National Bureau of Standards published special tables of Tchebycheff polynomials [16] (Russian transl., 1963).

151

An application of polynomials deviating least from zero on the point set formed by two equal intervals on the number line was obtained in fairly recent radio engineering investigations. These polynomials are denoted by A_n in certain American papers.[32] The precise definition of them is as follows.

Among all polynomials with leading term x^n the polynomial $A_n(x;\alpha) = x^n + \cdots$ deviates least from zero on the intervals

$$[-1, -\alpha], \qquad [\alpha, 1], \tag{1}$$

where α $(0 < \alpha < 1)$ is a given number; the deriviation of $A_n(x;\alpha)$ from zero is denoted by $L_n(\alpha)$.

The uniqueness of $A_n(x;\alpha)$ follows from simple considerations similar to those we used in §51. Therefore, it is even for even n and odd for odd n.

This gives us easily that the polynomial $A_{2m}(x;\alpha)$ can be expressed in a simple way in terms of the Tchebycheff polynomial $T_m(t)$. Indeed, if we set $x^2 = y$, then $A_{2m}(x;\alpha) = y^m + \cdots = P_m(y)$ deviates least from zero in $[\alpha^2, 1]$ among all polynomials in y with leading term y^m. We make the substitution

$$y = \frac{1-\alpha^2}{2} t + \frac{1+\alpha^2}{2}, \qquad P_m(y) = \frac{(1-\alpha^2)^m}{2^m} Q_m(t),$$

so that $Q_m(t) = t^m + \cdots$. As y runs through $[\alpha^2, 1]$, the new variable t runs through $[-1, 1]$.

Consequently, $Q_m(t)$ deviates least from zero in $[-1, 1]$ among all polynomials with leading term t^m, and hence $Q_m(t) = 2^{1-m} T_m(t)$.

On the basis of the foregoing,

$$A_{2m}(x;\alpha) = \frac{(1-\alpha^2)^m}{2^{2m-1}} T_m\left(\frac{2x^2 - 1 - \alpha^2}{1 - \alpha^2}\right),$$

$$L_{2m}(\alpha) = \frac{(1-\alpha^2)^m}{2^{2m-1}}.$$

If n is odd, then the matter is more complicated, with the exception of the case $n = 1$, of course, when the polynomial is trivial:

$$A_1(x;\alpha) = x, \qquad L_1(\alpha) = 1.$$

For each $n = 2m - 1 > 1$ the polynomial $A_{2m-1}(x;\alpha)$ is simply equal to $2^{2m-2} T_{2m-1}(x)$ for all sufficiently small α. Indeed, the points of deviation of $T_{2m-1}(x)$ are given by the formula

$$x_k = -\cos(k\pi/(2m - 1)) \qquad (k = 0, 1, \ldots, 2m - 1).$$

---

[32] They were first constructed by the author in 1928. See [19].

There are $2m$ of them: the points $x_0, \ldots, x_{m-1}$ belong to the interval $[-1, 0)$, and $x_m, \ldots, x_{2m-1}$ belong to $(0, 1]$. If $\alpha \le x_m$, then $x_{m-1} \le -\alpha$, and all the points x_k $(k = 0, 1, \ldots, 2m - 1)$ belong to the intervals in (1). Therefore, the polynomial $2^{2m-2} T_{2m-1}(x)$ is extremal on the system of intervals (1) when $\alpha \le x_m$, i.e., it coincides with $A_{2m-1}(x; \alpha)$.

But if

$$\alpha > x_m = -\cos \frac{m\pi}{2m - 1} = \sin \frac{\pi}{2(2m - 1)}, \tag{2}$$

then the extremal polynomial is different. Indeed, the number of points of deviation on the set (1) must be equal to $2m$ (in view of the general theorem of Tchebycheff), i.e., it must remain as before, but the deviation on $[-\alpha, \alpha]$ is now greater than on the intervals in (1).

Therefore, the polynomial will have a maximum at some point[33] γ of $(-\alpha, \alpha)$, and hence a minimum at the point $-\gamma$. Consequently, in $(-\alpha, \alpha)$ the polynomial $A_{2m-1}(x; \alpha)$ takes the value $L_{2m-1}(\alpha)$ at some point β, and the value $-L_{2m-1}(\alpha)$ at $-\beta$. If we introduce the points of deviation lying in the interval[34] $[\alpha, 1]$, namely,

$$\eta_0 = \alpha < \eta_1 < \eta_2 < \cdots < \eta_{m-1} = 1,$$

and for simplicity we write $A(x)$ instead of $A_{2m-1}(x; \alpha)$ and L instead of $L_{2m-1}(\alpha)$, then we arrive without difficulty at the relations

$$[A(x)]^2 - L^2 = (x^2 - \beta^2)(x^2 - \alpha^2)(x^2 - 1)(x^2 - \eta_1^2)^2 \cdots (x^2 - \eta_{m-1}^2)^2,$$

$$A'(x) = (2m - 1)(x^2 - \gamma^2)(x^2 - \eta_1^2) \cdots (x^2 - \eta_{m-1}^2).$$

They imply that

$$\frac{A'(x)}{\sqrt{L^2 - [A(x)]^2}} = \frac{(2m - 1)(x^2 - \gamma^2)}{\sqrt{(x^2 - \alpha^2)(x^2 - \beta^2)(1 - x^2)}}.$$

From this,

$$\arcsin \frac{A(x)}{L} = (2m - 1) \int_0^x \frac{(x^2 - \gamma^2)\, dx}{\sqrt{(x^2 - \alpha^2)(x^2 - \beta^2)(1 - x^2)}}. \tag{3}$$

Making the substitution $x^2 = t$, we get an elliptic integral on the right-hand side. Formula (3) contains the parameters β, γ, and L. They can be determined from the condition that $A_{2m-1}(x; \alpha)$ is a polynomial with leading term x^{2m-1}. Further, we expect that a parametric representation of the desired polynomials is obtained after the introduction of elliptic functions in (3).

All this is indeed true. However, we can arrive at the final result without long computations by using certain considerations in the geometric theory of functions.

Nevertheless, we do not dwell on this, but formulate and then prove the finished result.

---

[33]This point is unique, because otherwise the total number of points of deviation in the open interval $(\alpha, 1)$ would not be more than $m - 3$, and hence the total number of points of deviation on the set (1) would not be greater than $2(m - 3) + 4 = 2m - 2$.

[34]Now α must be a point of deviation.

This result is the following: *if $n = 2m - 1$ $(m > 1)$, $\alpha > \sin(\pi/2n)$ k $(0 < k < 1)$ is determined from the equation*

$$\operatorname{sn}(K/n; k) = \alpha \tag{4}$$

and

$$x = \frac{\alpha \operatorname{cn} u}{\sqrt{\alpha^2 - \operatorname{sn}^2 u}}, \tag{5}$$

where the radical is chosen so that $x = 1$ for $u = 0$, then

$$A_n(x; \alpha) = \frac{1}{2} L_n(\alpha) \left\{ \left[\frac{H(c+u)}{H(c-u)} \right]^{n/2} + \left[\frac{H(c-u)}{H(c+u)} \right]^{n/2} \right\}, \tag{6}$$

where $c = K/n$, and

$$L_n(\alpha) = \frac{1}{2^{n-1}} \left[\frac{\Theta(0)\Theta_1(0)}{\Theta(c)\Theta_1(c)} \right]^n. \tag{7}$$

Thus, (5) and (6) give a parametric representation of the polynomial, and (7) gives the magnitude of the deviation.

First of all it is necessary to show that for fixed $n > 1$ equation (4) is uniquely solvable with respect to k for an arbitrary α in the interval $(\sin(\pi/2n), 1)$.

With this goal we rewrite (4) in the form

$$\frac{1}{n} \int_0^1 \frac{dt}{\sqrt{(1 - t^2)(1 - k^2 t^2)}} = \int_0^\alpha \frac{dt}{\sqrt{(1 - t^2)(1 - k^2 t^2)}}.$$

From this, differentiating with respect to k, we get that

$$\frac{1}{n} \int_0^1 \frac{dt}{\sqrt{(1 - t^2)1 - k^2 t^2)}} \frac{kt^2}{1 - k^2 t^2}$$

$$= \frac{1}{\sqrt{(1 - \alpha^2)(1 - k^2 \alpha^2)}} \frac{d\alpha}{dk} + \int_0^\alpha \frac{dt}{\sqrt{(1 - t^2)(1 - k^2 t^2)}} \frac{kt^2}{1 - k^2 t^2}$$

or, in view of (4),

$$d\frac{d\alpha}{dk} = \frac{1}{n} \operatorname{cn} \frac{K}{n} \operatorname{dn} \frac{K}{n} \int_0^K \frac{du}{\operatorname{dn}^2 u} - \operatorname{cn} \frac{K}{n} \operatorname{dn} \frac{K}{n} \int^{K/n} \frac{dv}{\operatorname{dn}^2 v},$$

and hence

$$k\frac{d\alpha}{dk} = \frac{1}{n} \operatorname{cn} \frac{K}{n} \operatorname{dn} \frac{K}{n} \int_0^K \left[\frac{1}{\operatorname{dn}^2 u} - \frac{1}{\operatorname{dn}^2(u/n)} \right] du.$$

Since $n > 1$ and $0 < k < 1$, the right-hand side is positive. Therefore, as k increases from 0 to 1, the variable α increases monotonically from $\alpha = \sin(\pi/2n)$ to $\alpha = 1$. Consequently, for any α in the interval $(\sin(\pi/2n), 1)$ equation (4) is uniquely solvable with respect to k.

Passing to the proof of (6) and (7), we introduce the function

$$\varphi(u) = \frac{\sqrt{\operatorname{sn}^2 c - \operatorname{sn}^2 u}}{\operatorname{cn} u} \left\{ \left[\frac{H(c+u)}{H(c-u)} \right]^{n/2} + \left[\frac{H(c-u)}{H(c+u)} \right]^{n/2} \right\},$$

which can be reduced to the form

$$\varphi(u) = B\frac{[H(c+u)]^{2m-1} + [H(c-u)]^{2m-1}}{H_1(u)[H(c-u)H(c+u)]^{m-1}}, \tag{8}$$

where B is a constant.

It is immediately clear from (8) that $\varphi(-u) = \varphi(u)$, and with the help of the reduction formulas for theta functions it is easy to show that

$$\varphi(u + 2K) = \varphi(u + 2iK') = \varphi(u).$$

Thus, $\varphi(u)$ is an even meromorphic function of u with periods $2K$ and $2iK'$. Therefore, $\varphi(u)$ is a rational function of $\operatorname{sn}^2 u$, and hence a rational function of

$$x^2 = \frac{\alpha^2(1 - \operatorname{sn}^2 u)}{\alpha^2 - \operatorname{sn}^2 u}.$$

In the period parallelogram

$$-K < \Re u \leq K, \qquad 0 \leq \Im u < 2K'$$

$H_1(u)$ vanishes only at $u = K$. However, the numerator of the right-hand side of (8) also vanishes there. Therefore, the poles $u = \pm c$ are the only singular points of $\varphi(u)$, and each of them has multiplicity $m - 1$. Since x^2 also has a pole, which is simple, at the points $u = \pm c$, it follows that $\varphi(u)$ is a polynomial in x^2, and it has degree $m - 1$. Hence, the right-hand side of (6) is an odd polynomial of degree $2m - 1$ in x.

A check that is not difficult and that we can omit shows that the leading term on the right-hand side of (6) is x^{2m-1} when the coefficient $L_n(\alpha)$ is chosen according to (7).

Now observe that the segment of the imaginary axis in the u-plane from the point iK' to the point 0 is carried to the interval $[\alpha^2, 1]$ of the real axis of the x-plane with the help of (5). Therefore,

$$|H(c + u)/H(c - u)| = 1$$

for $\alpha \leq x \leq 1$, and hence

$$|A_n(x; \alpha)| \leq L_n(\alpha).$$

At the points of the interval $\alpha \leq x \leq 1$ where

$$\frac{n}{2}\arg\frac{H(c+u)}{H(c-u)} = r\pi$$

and r is an integer, we have the equality

$$A_n(x; \alpha) = \pm L_n(\alpha).$$

It remains for us to trace the change in the quantity

$$\arg \frac{H(c+u)}{H(c-u)} \tag{9}$$

when u runs through the segment of the imaginary axis from $u = 0$ to $u = iK'$. We assume that (9) is equal to zero at $u = 0$. By the reduction formulas for theta functions,

$$\frac{H(c+iK')}{H(c-iK')} \equiv \frac{H(K/n+iK')}{H(K/n-iK')} = -e^{-\pi i/n}.$$

Therefore, at the point $u = iK'$ the quantity (9) is numerically greater than or equal to $\pi(1 - 1/n) = 2(m-1)\pi/n$. Consequently, the polynomial (6) takes its maximum on $[\alpha, 1]$ with alternating signs at least m times,[35] and is thus an extremal polynomial.

This proves our assertion.

§53. Orthogonal polynomials on two intervals

The Tchebycheff polynomials form an orthogonal system on $[-1, 1]$ with respect to the weight $(1 - t^2)^{-1/2}/\pi$:

$$\frac{1}{\pi} \int_{-1}^{1} T_m(t) T_n(t) \frac{dt}{\sqrt{1-t^2}} = 0$$
$$(m \neq n; m, n = 0, 1, 2, \ldots),$$

as follows immediately from the definition in §52 of the Tchebycheff polynomials. Instead of $T_n(x)$ it is common to consider the polynomials

$$\widehat{T}_n(x) = \begin{cases} T_n(x), & (n = 0), \\ \sqrt{2}\,T_n(x) & (n \geq 1), \end{cases}$$

which are not only orthogonal, but also normalized:

$$\frac{1}{\pi} \int_{-1}^{1} \widehat{T}_m(t) \widehat{T}_n(t) \frac{dt}{\sqrt{1-t^2}} = \delta_{mn}.$$

In this section we occupy ourselves with the construction of orthogonal polynomials for the case when the domain of orthogonality is a pair of finite intervals, while the weight represents a generalization of the Tchebycheff weight.

We first prove an auxiliary proposition. Suppose that the weight is an arbitrary integrable function $p(t) \geq 0$ on the finite interval $[a, b]$. Let

----

[35] The polynomial (6) cannot have more than m points of deviation in $[\alpha, 1]$. Therefore, $2(m-1)\pi/n$ is the precise value of the quantity (9) at the point $u = iK'$.

$P_n(x)$ $(n \geq 0)$ be orthogonal polynomials such that $P_n(x)$ is a polynomial of degree precisely n, and the equalities

$$\int_a^b P_n(x)x^k p(x)\,dx = 0 \qquad (k = 0, 1, \ldots, n-1) \tag{1}$$

are satisfied for $n > 0$; these equalities determine it to within a constant factor.

We introduce for each $n \geq 1$ a polynomial $Q_n(x)$ of degree $n - 1$ according to the formula

$$Q_n(x) = \int_a^b \frac{P_n(x) - P_n(t)}{x - t} p(t)\,dt; \tag{2}$$

$Q_n(x)$ is called a *polynomial of the second kind*.

For example, the Tchebycheff polynomials of the second kind have the form

$$U_n(x) = \frac{\sin n\varphi}{\sin \varphi} = \frac{1}{n} T_n'(x) \qquad (x = \cos \varphi).$$

We introduce further the function

$$w(x) = \int_a^b \frac{p(t)}{x - t}\,dt, \tag{3}$$

which is analytic outside the interval $[a, b]$ of the real axis of the complex x-plane. By the definitions (2) and (3),

$$Q_n(x) = P_n(x)w(x) - \int_a^b \frac{P_n(t)}{x - t} p(t)\,dt.$$

From this we find that as $|x| \to \infty$

$$P_n(x)w(x) - Q_n(x) = O(1/x^{n+1}). \tag{4}$$

Indeed, due to (1),

$$\int_a^b P_n(t)\frac{x^n - t^n}{x - t} p(t)\,dt = 0,$$

and hence

$$\int_a^b \frac{P_n(t)}{x - t} p(t)\,dt = \frac{1}{x^n} \int_a^b \frac{t^n P_n(t)}{x - t} p(t)\,dt.$$

We now show that the relation (4) between $P_n(x)$ and $Q_n(x)$ is *characteristic* for an orthogonal polynomial and the corresponding polynomial of the second kind. This is the auxiliary proposition mentioned above.

Accordingly, suppose that the polynomial $P_n(x)$ of nth degree, the polynomial $Q_n(x)$ of $(n-1)$st degree, and the function $w(x)$ defined by (3) satisfy relation (4). It follows from this relation that

$$\left[\int_a^b \frac{P_n(x) - P_n(t)}{x - t} p(t)\, dt - Q_n(x)\right] + \int_a^b \frac{P_n(t)}{x - t} p(t)\, dt$$
$$= O\left(\frac{1}{x^{n+1}}\right).$$

Since the expression in the square brackets is a polynomial (of degree $\leq n-1$), while the remaining terms in the equality tend to zero as $|x| \to \infty$, the expression in the square brackets is identically zero, and hence $Q_n(x)$ and $P_n(x)$ are connected by relation (2), and, moreover,

$$\int_a^b \frac{P_n(t)}{x - t} p(t)\, dt = O\left(\frac{1}{x^{n+1}}\right),$$

which implies that

$$\int_a^b P_n(t) \frac{t^n}{x - t} p(t)\, dt + \int_a^b P_n(t) \frac{x^n - t^n}{x - t} p(t)\, dt = O\left(\frac{1}{x}\right).$$

The second term on the left-hand side is a polynomial of degree $\leq n - 1$, and the remaining terms in the equality tend to zero at $|x| \to \infty$. Therefore, the second term on the left-hand side is identically zero, that is,

$$\sum_{k=0}^{n-1} x^{n-k-1} \int_a^b P_n(t) t^k p(t)\, dt = 0;$$

but this means that the relations (1) hold, and our proposition is proved, i.e., *if the polynomials $P_n(x)$ and $Q_n(x)$ satisfy relation (4), and $w(x)$ is defined by (3), then $P_n(x)$ is an orthogonal polynomial with respect to the weight $p(x)$, and $Q_n(x)$ is the corresponding polynomial of the second kind.*

Passing to our concrete case, we assume that the two given finite intervals of the number line are $[-1, \alpha]$ and $[\beta, 1]$ $(-1 < \alpha < \beta < 1)$. Let $\mathscr{E}$ denote the point set in the complex x-plane formed by these intervals. The weight is taken to be the function

$$p(t) = \begin{cases} \dfrac{1}{\pi} \sqrt{\dfrac{t - \alpha}{(1 - t^2)(t - \beta)}} \geq 0 & (t \in \mathscr{E}), \\ 0 & (\alpha < t < \beta) \end{cases}$$

(it passes into the Tchebycheff weight in the limit case $\alpha = \beta$). The function $w(x)$ is now analytic in the x-plane, cut along $\mathscr{E}$, and not only outside the interval $[-1, 1]$ of the real axis.

We remark that in the x-plane, cut along $\mathscr{E}$,

$$\int_{-1}^{1} \frac{p(t)}{x-t} \equiv \frac{1}{\pi} \int_{\mathscr{E}} \sqrt{\frac{t-\alpha}{(1-t^2)(t-\beta)}} \frac{dt}{x-t}$$

$$= \sqrt{\frac{x-\alpha}{(x^2-1)(x-\beta)}} \tag{5}$$

if the radical on the right-hand side is positive at some chosen real point $x > 1$, and then is extended by continuity.[36] Thus,

$$w(x) = \sqrt{\frac{x-\alpha}{(x^2-1)(x-\beta)}}.$$

In this case (4) takes the form

$$P_n(x)\sqrt{\frac{x-\alpha}{(x^2-1)(x-\beta)}} - Q_n(x) = O\left(\frac{1}{x^{n+1}}\right), \tag{6}$$

and since, by our convention about the radical,

$$P_n(x)\sqrt{\frac{x-\alpha}{(x^2-1)(x-\beta)}} + Q_n(x) = O(x^{n-1}),$$

it follows that

$$\frac{x-\alpha}{(x^2-1)(x-\beta)} P_n^2(x) - Q_n^2(x) = O\left(\frac{1}{x^2}\right),$$

or

$$(x-\alpha)P_n^2(x) + (1+x)(x-\beta)(1-x)Q_n^2(x) = O(x).$$

Here the left-hand side is a polynomial; consequently, the right-hand side is equal to $Ax + B$, where A and B are constants (depending, of course, on n). Therefore,

$$P_n^2(x) + \frac{(1+x)(x-\beta)(1-x)}{x-\alpha} Q_n^2(x) = \frac{Ax+B}{x-\alpha}.$$

---

[36] To prove (5) it is necessary to apply to the function

$$\sqrt{\frac{x-\alpha}{(x^2-1)(x-\beta)}}$$

the Cauchy integral formula, taking the triply connected domain bounded by a circle of infinite radius and two oval curves encircling the segments $[-1, \alpha]$ and $[\beta, 1]$, and then to shrink the ovals to the indicated segments.
Incidentally, (5) implies that

$$\frac{1}{\pi} \int_{\mathscr{E}} \sqrt{\frac{t-\alpha}{(1-t^2)(t-\beta)}} \, dt = 1.$$

This is an indeterminate equation, and by solving it in polynomials we must obtain the orthogonal polynomials $P_n(x)$ and the polynomials of the second kind $Q_n(x)$. We show how to do this with the help of elliptic functions. With this goal we construct first of all a two-sheeted Riemann surface $\mathfrak{F}$ with branch points -1, α, β, and 1 and with transition lines $[-1, \alpha]$ and $[\beta, 1]$. The function $w(x)$ is single-valued on this surface $\mathfrak{F}$. On the upper sheet $\mathfrak{F}^+$ it coincides with the function defined earlier in the cut plane. On the lower sheet $\mathfrak{F}^-$ it differs only in sign (at corresponding points). We now introduce on $\mathfrak{F}$ the function

$$\mathscr{E}(x) = P_n(x) - \frac{1}{w(x)} Q_n(x)$$

$$= \frac{Ax + B}{(x - \alpha)\{P_n(x) + (1/w(x))Q_n(x)\}}. \tag{7}$$

In view of (6), $\mathscr{E}(x)$ *has a root of multiplicity n at the point at infinity in the sheet $\mathfrak{F}^+$, and hence* (by virtue of the second representation) *a pole of multiplicity n at the point at infinity in the sheet $\mathfrak{F}^-$;* moreover, $\mathscr{E}(x)$ *has a simple pole at the point $x = \alpha$,* as is clear from its first representation, *and another simple root* at some as yet unknown point. (This is clear from the second representation).

In §49 (Example 3) we considered a conformal mapping of a rectangle in the u-plane onto the plane cut along $\mathscr{E}$ (now the sheet $\mathfrak{F}^-$ is this plane).

The indicated rectangle Δ^- is given by the inequalities

$$-K \le \mathfrak{R}u \le 0, \qquad -K' \le \mathfrak{J}u \le K',$$

and the modulus k $(0 < k < 1)$ is equal to

$$k = \sqrt{\frac{2(\beta - \alpha)}{(1 - \alpha)(1 + \beta)}}, \tag{8}$$

while the mapping is given by

$$x = \frac{\mathrm{sn}^2 u \, \mathrm{cn}^2 \rho + \mathrm{cn}^2 u \, \mathrm{sn}^2 \rho}{\mathrm{sn}^2 u - \mathrm{sn}^2 \rho}, \tag{9}$$

where the parameter ρ $(0 < \rho < K)$ is determined from

$$1 - 2 \, \mathrm{sn}^2 \rho = \alpha. \tag{10}$$

Thus, $u = -\rho$ is the image of the point at infinity in the sheet $\mathfrak{F}^-$. The same formulas map the rectangle Δ^+ symmetric with respect to the imaginary axis onto the sheet $\mathfrak{F}^+$. Consequently, $u = \rho$ is the image of the point at infinity in $\mathfrak{F}^+$.

The whole Riemann surface $\mathfrak{F}$ is mapped onto the rectangle

$$-K \leq \Re u \leq K, \qquad -K' \leq \Im u \leq K',$$

with its opposite sides necessarily identified.

It is not hard to derive the following formulas from (8)–(10):

$$x - \alpha = \frac{1 - \alpha^2}{2(\operatorname{sn}^2 u - \operatorname{sn}^2 \rho)},$$

$$w(x) = \sqrt{\frac{(x^2 - 1)(x - \beta)}{x - \alpha}}$$

$$= \frac{2 \operatorname{sn} \rho \operatorname{cn} \rho}{\operatorname{dn} \rho} \frac{\operatorname{sn} u \operatorname{cn} u \operatorname{dn} u}{\operatorname{sn}^2 u - \operatorname{sn}^2 \rho}.$$

We see that $w(x)$, as should be expected, is a function with periods $2K$ and $2iK'$, and also that changing the sign of u corresponds to passing from a point of one sheet of $\mathfrak{F}$ to the corresponding point of the other sheet.

The function $\mathscr{E}(x)$ is elliptic with periods $2K$ and $2iK'$. It follows from the italicized properties of it on the surface $\mathfrak{F}$ that in a period rectangle $\mathscr{E}(x)$ has an n-fold root at the point $u = \rho$, an n-fold pole at the point $u = -\rho$, a simple pole at the point $u = iK'$ (since $x - \alpha$ vanishes for $u = iK'$), and another simple root at some point u_0. By Liouville's theorem,

$$-n\rho + iK' \equiv n\rho + u_0 \ [\mathrm{mod}(2K, 2iK')],$$

it follows that

$$\mathscr{E}(x) = C \left[\frac{H(u - \rho)}{H(u + \rho)} \right]^n \frac{H(u + 2n\rho - iK')}{H(u - iK')},$$

or, with another constant C,

$$\mathscr{E}(x) = C \left[\frac{H(u - \rho)}{H(u + \rho)} \right]^n \frac{\Theta(u + 2n\rho)}{\Theta(u)}.$$

Using the definition (7) of the function $\mathscr{E}(x)$, we find that

$$P_n(x) + \frac{1}{w(x)} Q_n(x) = C \left[\frac{H(u + \rho)}{H(u - \rho)} \right]^n \frac{\Theta(u - 2n\rho)}{\Theta(u)}.$$

Therefore,

$$P_n(x) = \frac{C_n}{2} \left\{ \left[\frac{H(u - \rho)}{H(u + \rho)} \right]^n \frac{\Theta(u + 2n\rho)}{\Theta(u)} \right.$$

$$\left. + \left[\frac{H(u + \rho)}{H(u - \rho)} \right]^n \frac{\Theta(u - 2n\rho)}{\Theta(u)} \right\}.$$

This formula together with (9), (8), and (10) gives a parametric represen-
ation of the orthogonal polynomials on a pair of intervals for the weight
considered.[37] The normalization condition leads to the value

$$C_n = \frac{\sqrt{2}\Theta(\rho)}{\sqrt{\Theta((2n-1)\rho)\Theta((2n+1)\rho)}}.$$

---

[37] Concerning the constructions in this section see the author's papers [25] and [26].
There is a generalization to the case of n intervals in [27]–[29].

CHAPTER 11

Various Supplements and Applications

§54. Abel's theorem

Suppose that we are given an irreducible algebraic equation, i.e., an equation of the form

$$F(z, w) = 0, \qquad (1)$$

where $F(z, w)$ is a polynomial in z and w that is not representable as a product of two analogous (nonconstant) polynomials. The set of all pairs (z, w) satisfying (1) is called an *algebraic curve*, and each pair (z, w) is a *point* of this curve. The algebraic function $w = w(z)$ determined by (1) is not single-valued, except in the case when (1) is linear in w. However, w can be made a single-valued point function if instead of the complex plane we take a multisheeted surface—the Riemann surface belonging to equation (1). On this surface all rational functions $R(z, w)$ of z and w will be single-valued simultaneously with w.

If we do not resort to this Riemann surface but remain in the complex plane, then to study the function $w(z)$ (as well as the functions $R(z, w)$) we make certain cuts in the z-plane and consider each branch of $w(z)$ or $R(z, w)$ separately.

An especially simple picture emerges in the case when (1) has the form

$$w^2 = A(z - \alpha_1)(z - \alpha_2) \cdots (z - \alpha_{2p+2}), \qquad (2)$$

or

$$w^2 = A(z - \alpha_1)(z - \alpha_2) \cdots (z - \alpha_{2p+1}), \qquad (2^{\text{bis}})$$

where the α_i (the critical points) are distinct. If $p = 1$, we get an elliptic curve,[38] and for $p > 1$ a hyperelliptic curve.

Here the Riemann surface consists of two sheets. The branch points are the points α_i, with the point $z = \infty$ adjoined in the case (2bis). The transition lines can be chosen in various ways, just as long as they do not

---

[38] This has already been defined and considered in §41.

163

intersect. For example, in the case of equation (2) we can take the arcs $\alpha_{2k-1}\alpha_{2k}\cdots$ $(k = 1,\ldots,p+1)$ as transition lines. The plane with cuts along these arcs can be used if we study each of the two branches of the function w separately.

In the case when the function w is determined by equation (1) in general form, its finite critical points are the roots of the equation $D(z) = 0\,(^{39})$ obtained by eliminating w from (1) and the equation $\partial F(z,w)/\partial w = 0$.

As we know, integrals of the form

$$\int R(z,w)\,dz \tag{3}$$

based on (2) or (2bis) with $p = 1$ are called elliptic integrals. For $p > 1$ they bear the name *hyperelliptic integrals*.

Elliptic and hyperelliptic integrals, as well as the elementary integrals.

$$\int R(z,\sqrt{az^2 + bz + c})\,dz$$

are special cases of Abelian integrals, understood in general to be integrals of the form (3) with w and z connected by an irreducible algebraic equation (1).

The value of an Abelian integral, taken along some curve, depends not only on the initial point and endpoint, but also on the path of integration joining them. If the upper limit of the integral is made a variable, the integral will thus be a multivalued function of the upper limit.

For example, we have seen that the different values of the integral

$$\int \frac{dz}{\sqrt{4z^3 - g_2 z - g_3}}$$

have the form $\pm u + 2m\omega + 2m'\omega'$, where u is one of these values, m and m' are arbitrary integers, and ω and ω' are half-periods.

The Abelian integrals (3) belonging to the general equation (1) break up into three groups. This partition is analogous to that made above with elliptic integrals.

The following principle lies at the basis of the classification of Abelian integrals: an integral is called an *integral of the first kind* if it is finite everywhere, an *elementary integral of the second kind* if it becomes infinite at a single point, with the singularity algebraic, and, finally, an *elementary integral of the third kind* if it becomes infinite logarithmically at two points.

Recalling §§17 and 29, the reader can easily verify that the elliptic integrals of different kinds really are characterized only by these properties.

$(^{39})$The polynomial $D(z)$ is the discriminant of equation (1).

We cannot concern ourselves here with various questions in the extensive theory of Abelian integrals. However, there is one theorem that is very important with regard to applications, and at the same time simple enough in its classical variant that it must be presented even in an elementary course in elliptic functions. This is Abel's theorem.

Proceeding to this theorem, we take an irreducible equation (1) determining an algebraic function $w(z)$. Moreover, suppose that we are given a second algebraic equation of the form (1) whose left-hand side depends rationally on some parameters $a_1, \ldots, a_\mu$. To avoid additional purely algebraic considerations we confine ourselves to the case when this second equation has the form

$$w = g(z, a_1, \ldots, a_\mu), \tag{4}$$

where the right-hand side is a rational function of its arguments.

Eliminating w from (1) and (4), we get the equation

$$\varphi(z) = 0, \tag{5}$$

where the left-hand side is a polynomial whose coefficients depend rationally on the parameters a_i.

Aside from values of the a_i that are connected by certain algebraic relations, equation (5) has only simple roots. Denote them by $z_1, \ldots, z_N$. With the help of (4) we find the corresponding values of w: $w_1, \ldots, w_N$. Then we form all the pairs (z_i, w_i) $(i = 1, \ldots, N)$; we have thereby found all the intersection points of the algebraic curves (1) and (4).

Note that, for an arbitrary rational function $S(z, w)$ of its arguments that also depends rationally on the parameters a_i, the sum $S(z_1, w_1) + \cdots + S(z_N, w_N)$ is a rational function of the parameters a_i by virtue of a well-known theorem in higher algebra.

We now take a point (z_0, w_0) on the algebraic curve (1) and consider the Abelian integral

$$\int_{(z_0, w_0)}^{(z_k w_k)} R(z, w)\, dz, \tag{6}$$

taken in the z-plane from z_0 to z_k along some path that carries w continuously from the initial value w_0 to the final value w_k. The sum of all the integrals (6) is denoted by V:

$$V = \sum_{k=1}^{N} \int_{(z_0, w_0)}^{(z_k, w_k)} R(z, w)\, dz.$$

This sum clearly depends on the values of the parameters a_i. If they vary continuously and, moreover, in such a way that the changing points

$z_1, \ldots, z_N$ do not go through critical points of the function w nor through points where $R(z, w)$ is infinite, then V also varies continuously.

Abel's theorem asserts that V *is equal to a rational function of the parameters $a_1, \ldots, a_\mu$, added to a sum of logarithms of similar functions multiplied by constants.*

To prove this theorem we let the parameters $a_1, \ldots, a_\mu$ vary and find the differential of V. It is equal to

$$dV = \sum_{k=1}^{N} R(z_k, w_k) \left(\frac{\partial z_k}{\partial a_1} \, da_1 + \cdots + \frac{\partial z_k}{\partial a_\mu} \, da_\mu \right).$$

On the other hand, differentiating the identity $\varphi(z_k) = 0$ with respect to a_λ, we get that

$$\frac{\partial \varphi(z_k)}{\partial z_k} \frac{\partial z_k}{\partial a_\lambda} + \frac{\partial \varphi(z_k)}{\partial a_\lambda} = 0,$$

which implies that

$$\frac{\partial z_k}{\partial a_\lambda} = -\frac{\partial \varphi(z_k)/\partial a_\lambda}{\partial \varphi(z_k)/\partial z_k} = \rho\lambda(z_k),$$

where $\rho_\lambda(z_k)$ is a rational function of z_k and the parameters $a_1, \ldots, a_\mu$.

We can now represent dV in the form

$$dV = \sum_{\lambda=1}^{\mu} da_\lambda \left\{ \sum_{k=1}^{N} R(z_k, w_k) \rho_\lambda(z_k) \right\}.$$

The quantity

$$\sum_{k=1}^{N} R(z_k, w_k) \rho_\lambda(z_k)$$

is a symmetric rational function of the pairs (z_k, w_k) whose coefficients are rational functions of the parameters $a_1, \ldots, a_\mu$. Consequently, this quantity is a rational function of $a_1, \ldots, a_\mu$:

$$\sum_{k=1}^{N} R(z_k, w_k) \rho_\lambda(z_k) = R_\lambda(a_1, \ldots, a_\mu).$$

To get V it is necessary to integrate the total differential

$$\sum_{\lambda=1}^{\mu} R_\lambda(a_1, \ldots, a_\mu) \, da_\lambda.$$

This integration leads to a rational function of the parameters, added to logarithms of such functions multiplied by certain constants. Thus, Abel's theorem is proved.

In the especially interesting and important case when (6) is an integral of the first kind, the rational-logarithmic function of the parameters a_i must become a constant, and hence V is equal to a constant.

Indeed, otherwise there would be values of the parameters a_i for which V would be infinite, which is impossible, since integrals of the first kind are always finite.

Abel's theorem and this remark can be used to derive the addition theorem for the function $\wp$. Here the role of (1) is played by the equation

$$w^2 = 4z^3 - g_2 z - g_3.$$

As (4) we take the equation

$$w = a_1 z - a_2. \tag{4'}$$

Finally, suppose that the integral is $\int w^{-1}\, dz$. Equation (5) takes the form

$$\varphi(z) \equiv 4z^3 - g_2 z - g_3 - (a_1 z + a_2)^2 = 0. \tag{5'}$$

Let z_1, z_2, and z_3 be the roots of this equation, assumed to be simple, and let w_1, w_2, and w_3 be the corresponding values of w. In this case Abel's theorem and the remark give us that

$$\int_\infty^{(z_1, w_1)} \frac{dz}{w} + \int_\infty^{(z_2, w_2)} \frac{dz}{w} + \int_\infty^{(z_3, w_3)} \frac{dz}{w} = C = \text{const}.$$

If we move the line (4') to infinity, then the upper limit in each of the integrals becomes ∞. This means that all three integrals become integrals along closed contours, i.e., each of them is equal to some period. Therefore, $C = 2m\omega + 2m'\omega'$, and so

$$\int_\infty^{(z_1, w_1)} \frac{dz}{w} + \int_\infty^{(z_2, w_2)} \frac{dz}{w} + \int_\infty^{(z_3, w_3)} \frac{dz}{w} = 2m\omega + 2m'\omega'.$$

Let([40])

$$z_1 = \wp(u_1), \qquad z_2 = \wp(u_2), \qquad z_3 = \wp(u_3),$$
$$w_1 = \wp'(u_1), \qquad w_2 = \wp'(u_2), \qquad w_3 = \wp'(u_3).$$

Then our result can be formulated as follows: if

$$u_1 + u_2 + u_3 \equiv 0 \pmod{2\omega, 2\omega'}, \tag{7}$$

then there exist a_1 and a_2 such that

$$\wp'(u_1) = a_1 \wp(u_1) + a_2,$$
$$\wp'(u_2) = a_1 \wp(u_2) + a_2,$$
$$\wp'(u_3) = a_1 \wp(u_3) + a_2;$$

([40])It is assumed that none of the u_k is congruent to zero modulo the periods.

or, in other words, (7) implies that

$$\begin{vmatrix} 1 & \wp(u_1) & \wp'(u_1) \\ 1 & \wp(u_2) & \wp'(u_2) \\ 1 & \wp(u_3) & \wp'(u_3) \end{vmatrix} = 0.$$

This is one form of the addition theorem for the function $\wp$.

It can be shown that other addition theorems (for example, the addition theorem for the function ζ) are also included in Abel's theorem as special cases.

Of course, this theorem takes on special significance when we pass to more complicated Abelian integrals—first of all, when we pass to hyperelliptic integrals.

Here we confine ourselves to a single result obtained with the help of Abel's theorem.

Suppose that a hyperelliptic curve is being considered. Every rational function on it clearly has the form

$$G(z, w) = \frac{A(z)w + B(z)}{C(z)w + D(z)},$$

where A, B, C, and D are polynomials. To obtain the points on the hyperelliptic curve where $G(z, w)$ takes some value a it is necessary to adjoin the equation

$$G(z, w) = a \tag{8}$$

to (2). But this last equation can be rewritten in the form[41]

$$w = g(z, a), \tag{4''}$$

where the right-hand side is a rational function of z and a.

We take two values of the parameter a: a' and a''. Let

$$P'_k = (z'_k, w'_k) \qquad (k = 1, \ldots, N)$$

be the points of intersection of the curves (2) and (4'') for $a = a'$, and, similarly, let

$$P''_k = (z''_k, w''_k) \qquad (k = 1, \ldots, N)$$

be the points of intersection of the curves (2) and (4'') for $a = a''$.

We take some integral of the first kind

$$\int_{(z_0, w_0)}^{(z, w)} R(z, w)\, dz = \Phi(P),$$

---

[41]We assume here that

$$\begin{vmatrix} A(z) & B(z) \\ C(z) & D(z) \end{vmatrix} \not\equiv 0.$$

We do not dwell on the case when this condition is not satisfied.

where $P = (z, w)$, and we consider the sums

$$V' = \sum_{k=1}^{N} \Phi(P'_k), \qquad V'' = \sum_{k=1}^{N} \Phi(P''_k).$$

Each of these sums is equal to some constant. Therefore,

$$\sum_{k=1}^{N} \{\Phi(P'_k) - \Phi(P''_k)\} = C.$$

To compute this constant C we vary the upper limits in the integrals

$$\Phi(P'_k) = \int_{(z_0, w_0)}^{(z'_k, w'_k)} R(z, w)\, dz \qquad (k = 1, \ldots, N)$$

by continuously varying the parameter a from a' to a''. As a result of this procedure each difference $\Phi(P'_k) - \Phi(P''_k)$ becomes an integral along a closed contour,

$$\oint_{\Gamma_k} R(z, w)\, dz = \oint_{\Gamma_k} d\Phi.$$

Therefore,

$$C = \oint_{\sum_1^N \Gamma_k} d\Phi = \oint_{\Gamma} d\Phi$$

is also the integral of $d\Phi$ along some "multiple" closed contour, and this contour does not depend on the choice of the integral Φ of the first kind, while $\oint_{\Gamma} d\Phi$ represents some *period* of Φ.

Our result can be formulated as follows:[42] *if G is a rational function on a hyperelliptic curve and Φ is an arbitrary integral of the first kind, then the sum of the values of Φ at the points of the curve where $G = a'$ is congruent to the sum of the values of Φ at the points where $G = a''$ modulo the periods of Φ.*

It is easy to see that this is a generalization of one of Liouville's theorems on elliptic functions proved in §4.

§55. The Green's function for a circular annulus

In the w-plane let G be the circular annulus determined by the circles $|w| = 1$ and $|w| = h$, where the given positive number h is less than 1, and let c be a point in the interior of G; without loss of generality c can be assumed to lie on the positive real axis, so that $h < c < 1$.

We pose the problem of finding an analytic function $f(w)$ satisfying the following conditions:

---

[42] The result is true for arbitrary algebraic curves, and not only hyperelliptic curves.

a) $f(w)$ is regular inside and on the boundaries of G.

b) $|f(w)|$ is a single-valued function inside G.

c) $|f(w)|$ is equal to 1 on the boundaries of G.

d) $f(w)$ has a simple zero at $w = c$, and does not vanish at other points of G.

It is easy to see that

$$\Re \ln f(w) = \ln |f(w)|$$

is none other than the Green's function for the circular annulus G. Therefore, $f(w)$ certainly exists. Moreover, $f(w)$ is uniquely determined up to an arbitrary constant factor of modulus 1.

The function $f(w)$ is sometimes called the complex Green's function for the annulus under consideration.

Let $w_0 = 1/\overline{w}$ and $w_1 = h^2/\overline{w}$. The point w_0 is the mirror reflection of w with respect to the outer boundary of the annulus G, and w_1 is the mirror reflection with respect to the inner boundary.

Since $f(w)$ has modulus 1 on each of the boundaries of G, its values at w_0 and w_1 are connected with the value at w by the relations

$$f(w)\overline{f(w_0)} = 1, \qquad f(w)\overline{f(w_1)} = 1. \tag{1}$$

The function $f(w)$ can be continued analytically by these relations beyond the boundaries of G to two annuli adjacent to G, and then to two further annuli, and so on, and in the limit to the whole w-plane, punctured at the points $w = 0$ and $w = \infty$.

Since $f(w)$ has a simple zero at the point c of G, $f(w)$ has by virtue of (1) the simple poles $w = 1/c$ and $w = h^2/c$. Consequently, again by virtue of (1), $f(w)$ has the simple zeros $w = c/h^2$ and $w = h^2 c$. These considerations show that $f(w)$ has simple zeros

$$w = h^{2k} c \qquad (k = 0, \pm 1, \pm 2, \dots)$$

and simple poles

$$w = h^{2k}/c \qquad (k = 0, \pm 1, \pm 2, \dots),$$

and also that $f(w)$ does not have other zeros nor poles.

On the basis of the foregoing it is natural to consider the function

$$F(w) = \frac{(1 - \frac{w}{c}) \prod_{k=1}^{\infty}(1 - h^{2k}\frac{w}{c})(1 - h^{2k}\frac{c}{w})}{(1 - cw) \prod_{k=1}^{\infty}(1 - h^{2k}cw)(1 - h^{2k}\frac{1}{cw})},$$

which has the same zeros and the same poles as $f(w)$. Replacing w by w_0 and by w_1, we get that

$$\overline{F(w_0)} = 1/c^2 F(w), \qquad \overline{F(w_1)} = 1/F(w). \tag{2}$$

Therefore, $F(w)$ does not satisfy (1), and hence does not satisfy the requirement c).

However, we show that there are real constants λ and μ such that the function

$$\mu w^\lambda F(w) = F_1(w)$$

does satisfy c). Indeed, writing condition (1) for $F_1(w)$, we get that

$$\mu w^\lambda F(w)\overline{\mu w_0^\lambda}\,\overline{F(w_0)} = 1,$$
$$\mu w^\lambda F(w)\overline{\mu w_1^\lambda}\,\overline{F(w_1)} = 1.$$

On the basis of (2), these equalities take the form $\mu^2 = c^2$ and $\mu^2 h^{2\lambda} = 1$, which implies that $\mu = \pm c$ and $\lambda = -(\ln c)/\ln h$.

Therefore, the function

$$\pm c w^{-\ln c/\ln h} F(w)$$

satisfies all the requirements of the problem. It is not hard to express it in terms of theta functions. The data needed for this is contained in Table IX.

The final expression for $f(w)$ has the form

$$f(w) = \pm w^{-\ln c/\ln h}\vartheta_1\left(\left.\frac{\ln w - \ln c}{2\pi i}\right|\tau\right)\bigg/\vartheta_1\left(\left.\frac{\ln w + \ln c}{2\pi i}\right|\tau\right), \quad (3)$$

where $\tau = (\ln h)/\pi i$. Since a solution has been obtained, we can introduce $e^{i\delta}$ instead of the factor ± 1, where δ is an arbitrary real number.

Another expression of $f(w)$ is sometimes useful, namely, that obtained by passing from τ to $-1/\tau$. On the basis of the formulas in Table XVII we find without difficulty that

$$f(w) = \pm\vartheta_1\left(\left.\frac{\ln w - \ln c}{2\pi i\tau}\right| -\frac{1}{\tau}\right)\bigg/\vartheta\left(\left|\frac{\ln w + \ln c}{2\pi i\tau}\right| -\frac{1}{\tau}\right). \quad (4)$$

§56. The Dirichlet problem for a circular annulus

Let G be the circular annulus bounded by the circles $|w| = 1$ and $|w| = h$ in the w-plane, where the given positive number h is less than 1. It is required to find a function $F(w)$ that is regular and single-valued interior to G if the values of its real part on the boundaries are known.

The analogous problem for the case of the disk is solved by the Schwarz formula

$$F(w) = \frac{1}{2\pi}\int_0^{2\pi}\frac{1 + we^{1-is}}{1 - we^{-is}}\Phi(s)\,ds + iC. \quad (1)$$

Here it is assumed that the radius of the disk is equal to 1, and the position of the point on the circle is determined by the argument s of this point, so that $\Phi(s)$ represents the value of the real part of the desired function at the point e^{is}. As for C, it is an arbitrary real constant.

The Schwarz formula is derived and studied[43] in courses in the theory of analytic functions under sufficiently general assumptions about $\Phi(s)$.

Our problem is to pass from the disk to the annulus and to construct a formula analogous to (1).

Denote by $\Phi(s)$ and $\varphi(s)$ the values of the real part of $F(w)$ at points with argument s on the outer and inner boundaries, respectively, of G. We do not need to make very general assumptions about $\Phi(s)$ and $\varphi(s)$, because our main goal is to determine how the formula is affected by passing from a simply connected domain to a doubly connected one. It suffices to assume that $\Phi(s)$ and $\varphi(s)$ are piecewise continuous.

The quantity

$$\frac{1}{2\pi} \int_0^{2\pi} F(re^{is})\, ds = \frac{1}{2\pi i} \oint \frac{F(w)}{w}\, dw,$$

where the integral on the right-hand side is taken over the circle of radius r ($h < r < 1$) about the point $w = 0$, clearly does not depend on r. The real part of the integral has the same property. From this, letting r go first to 1 and then to h, and observing that we can take the required limits in

---

(43) The starting point of this study is the Poisson formula

$$U(r, s_0) \equiv \Re F(re^{is_0})$$

$$= \frac{1}{2\pi} \int_0^{2\pi} \frac{1 - r^2}{1 - 2r\cos(s - s_0) + r^2}\Phi(s)\, ds, \qquad (1^{\text{bis}})$$

which follows from the Schwarz formula (1). Since

$$\frac{1}{2\pi} \int_0^{2\pi} \frac{1 - r^2}{1 - 2r\cos(s - s_0) + r^2}\, ds = 1 \qquad (0 \le r < 1),$$

it follows that

$$U(r, s_0) - \Phi(s_0) = \frac{1}{2\pi} \int_0^{2\pi} \frac{1 - r^2}{1 - 2r\cos(s - s_0) + r^2}[\Phi(s) - \Phi(s_0)]\, ds$$

$$= \frac{1}{2\pi} \int_{-\delta}^{\delta} \frac{1 - r^2}{1 - 2r\cos t + r^2}[\Phi(s_0 + t) - \Phi(s_0)]\, dt$$

$$+ \frac{1}{2\pi} \int_{\delta \le |t| \le \pi} = I_1 + I_2.$$

If the absolutely integrable function $\Phi(s)$ is continuous at s_0, then it is easy to derive from this that

$$\lim_{r \to 1} U(r, s_0) = \Phi(s_0).$$

Indeed, on the one hand, $I_1 < \varepsilon$ in modulus for all r ($0 \le r < 1$) if δ is sufficiently small (because $\Phi(s)$ is continuous), and, on the other hand, for fixed $\delta > 0$ the quantity I_2 tends to 0 as $r \to 1$ (because $\Phi(s)$ is absolutely integrable).

the integral

$$\frac{1}{2\pi} \int_0^{2\pi} \Re F(re^{is})\,ds,$$

we get that

$$\int_0^{2\pi} \Phi(s)\,ds = \int_0^{2\pi} \varphi(s)\,ds. \tag{2}$$

This condition is thus necessary for the solvability of the problem, and we must assume that it holds.

Suppose that the desired function $F(w)$ exists. In this case it can be expanded in a Laurent series:

$$F(w) = \sum_{k=-\infty}^{\infty} a_k w^k \qquad (h < |w| < 1). \tag{3}$$

The coefficients of this series are determined by

$$a_k = \frac{1}{2\pi i} \oint \frac{F(w)}{w^{k+1}}\,dw \qquad (\pm k = 0, 1, 2, \dots),$$

where the integral can be taken over the circle $w = re^{is}$ $(h < r < 1)$. Therefore,

$$a_k r^k = \frac{1}{2\pi} \int_0^{2\pi} F(re^{is}) e^{-iks}\,ds \qquad (\pm k = 0, 1, 2, \dots).$$

From this,

$$\overline{a}_{-k} r^{-k} = \frac{1}{2\pi} \int_0^{2\pi} \overline{F(re^{is})} e^{-iks}\,ds.$$

Adding, we find that

$$a_k r^k + \overline{a}_{-k} r^{-k} = \frac{1}{\pi} \int_0^{2\pi} e^{-iks} \Re F(re^{is})\,ds.$$

We can pass to the limits as $r \to 1$ and as $r \to h$ in this equality, obtaining

$$a_k + \overline{a}_{-k} = \frac{1}{\pi} \int_0^{2\pi} \Phi(s) e^{-iks}\,ds,$$

$$a_k h^k + \overline{a}_{-k} h^{-k} = \frac{1}{\pi} \int_0^{2\pi} \varphi(s) e^{-iks}\,ds.$$

The coefficients a_k are obtained from this, each as a sum of two integrals. It now remains to substitute them in the right-hand side of (3) and take the sum. The following formula of Villat is obtained as a result:

$$F(w) = \frac{i\omega}{\pi^2} \int_0^{2\pi} \Phi(s)\zeta\left(\frac{\omega}{\pi i}\ln(w) - \frac{\omega}{\pi}s\right) ds$$

$$- \frac{i\omega}{\pi^2} \int_0^{2\pi} \varphi(s)\left[\zeta\left(\frac{\omega}{\pi i}\ln w - \frac{\omega}{\pi}s - \omega'\right) + \eta'\right] ds + iC, \tag{4}$$

where C is an arbitrary real constant, ω is an arbitrary positive number, the purely imaginary number ω' is found from the equality $e^{\pi i \omega'/\omega} = h$, and, finally,[44] $\zeta(u) = \zeta(u \mid \omega, \omega')$

Let us now prove that the function $F(w)$ defined by (4) satisfies all the requirements of the problem. We would not have been spared this proof even if we had performed the summation just mentioned and thereby given a derivation of (4), because the whole of this derivation was based on the assumption that a solution exists. This was the reason we thought we could omit the derivation.

The regularity of the function $F(w)$ defined by (4) interior to G follows from the fact that both the functions

$$\zeta\left(\frac{\omega}{\pi i} \ln w - \frac{\omega}{\pi} s\right), \qquad \zeta\left(\frac{\omega}{\pi i} \ln w - \frac{\omega}{\pi} s - \omega'\right)$$

are regular interior to G.

The single-valuedness is a consequence of relation (2). Indeed, to see that the function (4) is single-valued it is necessary to show that this function does not vary when the point w describes the circle $|w| = r$ $(h < r < 1)$. After such a circuit the right-hand side of (4) takes the form

$$\frac{i\omega}{\pi^2} \int_0^{2\pi} \Phi(s)\zeta\left(\frac{\omega}{\pi i} \ln w - \frac{\omega}{\pi} s + 2\omega\right) ds$$

$$- \frac{i\omega}{\pi^2} \int_0^{2\pi} \varphi(s)\left[\zeta\left(\frac{\omega}{\pi i} \ln w - \frac{\omega}{\pi} s - \omega' + 2\omega\right) + \eta'\right] ds + iC.$$

But in view of the relation $\zeta(u + 2\omega) = \zeta(u) + 2\eta$ this quantity is equal to

$$F(w) + \frac{2i\omega}{\pi^2}\eta\left\{\int_0^{2\pi} \Phi(s)\, ds - \int_0^{2\pi} \varphi(s)\, ds\right\}.$$

On the basis of (2) the integrals cancel each other, so that the expression obtained is $F(w)$, and single-valuedness is proved.

It remains to verify that the real part of the function (4) on the boundaries of G becomes the given functions.

Take a point $w_0 = e^{is_0}$ on the outer boundary of G and a point $w = re^{is_0}$ $(h < r < 1)$ with the same argument interior to G. We must prove that

$$\lim_{r \to 1} \Re F(re^{is_0}) = \Phi(s_0),$$

---

[44] We remark also (see Table V) that

$$\zeta\left(\tfrac{\omega}{\pi i} \ln w - \tfrac{\omega}{\pi} s - \omega'\right) + \eta' = \zeta_3\left(\tfrac{\omega}{\pi i} \ln w - \tfrac{\omega}{\pi} s\right).$$

provided only that $\Phi(s)$ is continuous at s_0. With this goal we rewrite (4) in the form

$$F(re^{is_0}) = \frac{1}{2\pi} \int_0^{2\pi} \Phi(s) \frac{e^{is} + re^{is_0}}{e^{is} - re^{is_0}} \, ds + iC$$

$$+ \frac{i\omega}{\pi^2} \int_0^{2\pi} \Phi(s) \left\{ \zeta \left(\frac{\omega}{\pi i} \ln r - \frac{\omega}{\pi} s + \frac{\omega}{\pi} s_0 \right) \right.$$

$$\left. - \frac{\pi i}{2\omega} \frac{re^{is_0} + e^{is}}{re^{is_0} - e^{is}} \right\} ds$$

$$- \frac{i\omega}{\pi^2} \int_0^{2\pi} \varphi(s) \left\{ \zeta \left(\frac{\omega}{\pi i} \ln r - \frac{\omega}{\pi} s + \frac{\omega}{\pi} s_0 - \omega' \right) + \eta' \right\} ds$$

$$= J_1 + iC + J_2 + J_3.$$

The expression in the curly brackets in the integral J_2 and the zeta function in J_3 are uniformly continuous functions of r $(h^{1/2} \le r \le 1)$ for $0 \le s \le 2\pi$. Therefore, an elementary theorem on passing to the limit (as $r \to 1$) is applicable to these two integrals:

$$\lim_{r \to 1} J_2 = \frac{i\omega}{\pi^2} \int_0^{2\pi} \Phi(s) \left\{ \zeta \left(\frac{\omega}{\pi} s_0 - \frac{\omega}{\pi} s \right) - \frac{\pi}{2\omega} \cot \frac{s_0 - s}{2} \right\} ds,$$

$$\lim_{r \to 1} J_3 = \frac{i\omega}{\pi^2} \int_0^{2\pi} \varphi(s) \zeta_3 \left(\frac{\omega}{\pi} s - \frac{\omega}{\pi} s_0 \right) ds.$$

Since ω and ω'/i are real, the integrands in both integrals are real, and hence $\lim_{r \to 1} J_2$ and $\lim_{r \to 1} J_3$ have purely imaginary values. Consequently, it remains to consider the behavior of the real part of the sum $J_1 + iC$ as $r \to 1$; but this sum coincides with the right-hand side of the Schwarz formula (1).

It is proved similarly that $\Re F(w)$ tends to $\varphi(s_0)$ if the point $w = re^{is_0}$ approaches he^{is_0}, and s_0 is a point of continuity of $\varphi(s)$.

Thus, the proof of Villat's formula is complete.

§57. Elliptic coordinates

Suppose that a, b, and c are positive numbers such that $a > b > c$. Denote by X, Y, Z the moving coordinates of a point, and by s a real parameter. Then

$$\frac{X^2}{a^2 - s} + \frac{Y^2}{b^2 - s} + \frac{Z^2}{c^2 - s} = 1$$

is the equation of a family of second-order surfaces confocal with the ellipsoid

$$\frac{X^2}{a^2} + \frac{Y^2}{b^2} + \frac{Z^2}{c^2} = 1. \tag{1}$$

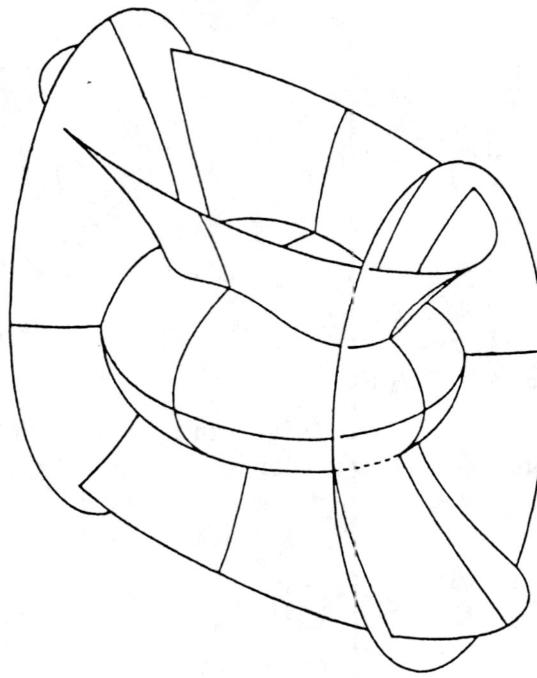

FIGURE 24

Here is it assumed, of course, that s is different from b^2 and c^2 and satisfies the inequality $s < a^2$.

Now let (x, y, z) be a point in space. Can a surface of our family be drawn through it? This question reduces to a question about the existence of real roots of the following equation:

$$f(s) \equiv 1 - \frac{x^2}{a^2 - s} - \frac{y^2}{b^2 - s} - \frac{z^2}{c^2 - s} = 0. \tag{2}$$

Consider the intervals

$$(-\infty, c^2 - \varepsilon), \qquad (c^2 + \varepsilon, b^2 - \varepsilon), \qquad (b^2 + \varepsilon, a^2 - \varepsilon).$$

For sufficiently small positive ε the function $f(s)$ is continuous in these intervals, and it has opposite signs at the endpoints of each of them. Consequently, equation (2) has a root in each of the intervals, i.e., all three of its roots are real. Denote them by λ, μ, and ν, and assume that $\lambda < \mu < \nu$.

We shall see that three surfaces of our family pass through each point (x, y, z) (Figure 24): an ellipsoid corresponds to the root λ, a hyperboloid of one sheet corresponds to μ, and, finally, a hyperboloid of two sheets corresponds to ν.

The quantities λ, μ, and ν can be regarded as the coordinates of a point. These coordinates are called *elliptic coordinates* with respect to the ellipsoid (1). It should be kept in mind that the correspondence between the Cartesian coordinates x, y, z and the elliptic coordinates λ, μ, ν is not one-to-one. Indeed, λ, μ, and ν are functions not of x, y, and z, but of x^2, y^2, and z^2, and hence the points $(\pm x, \pm y, \pm z)$ have the same elliptic coordinates λ, μ, ν as the point (x, y, z). However, if we consider not the whole space but only one octant, then in it the elliptic coordinates completely determine a point.

The coordinate surfaces passing through a given point are represented in Figure 24.

We now derive various relations and formulas used in applications of elliptic coordinates.

By the definition of λ, μ, and ν,

$$1 - \frac{x^2}{a^2 - s} - \frac{y^2}{b^2 - s} - \frac{z^2}{c^2 - s}$$
$$= -\frac{(s - \lambda)(s - \mu)(s - \nu)}{(a^2 - s)(b^2 - s)(c^2 - s)}.$$

From this, multiplying both sides by $a^2 - s$ and setting $s = a^2$, we have that

$$x^2 = \frac{(a^2 - \lambda)(a^2 - \mu)(a^2 - \nu)}{(a^2 - b^2)(a^2 - c^2)}. \tag{3_1}$$

Similarly,

$$y^2 = \frac{(b^2 - \lambda)(b^2 - \mu)(b^2 - \nu)}{(b^2 - c^2)(b^2 - a^2)},$$
$$z^2 = \frac{(c^2 - \lambda)(c^2 - \mu)(c^2 - \nu)}{(c^2 - a^2)(c^2 - b^2)}. \tag{3_2}$$

Let (x, y, z) be a point in space and let $\mathscr{L}$, $\mathscr{M}$, and $\mathscr{N}$ be the coordinate surfaces, passing through it. The direction cosines of the normals to these surfaces at the point under consideration are proportional to the following quantities:

$$\mathscr{L}: \quad x/(a^2 - \lambda), \ y/(b^2 - \lambda), \ z/(c^2 - \lambda),$$
$$\mathscr{M}: \quad x/(a^2 - \mu), \ y/(b^2 - \mu), \ z/(c^2 - \mu),$$
$$\mathscr{N}: \quad x/(a^2 - \nu), \ y/(b^2 - \nu), \ z/(c^2 - \nu).$$

We now observe that

$$\frac{x^2}{(a^2 - \lambda)(a^2 - \mu)} + \frac{y^2}{(b^2 - \lambda)(b^2 - \mu)} + \frac{z^2}{(c^2 - \lambda)(c^2 - \mu)}$$

$$= \frac{1}{\lambda - \mu} \left\{ \frac{x^2}{a^2 - \lambda} + \frac{y^2}{b^2 - \lambda} + \frac{z^2}{c^2 - \lambda} \right.$$

$$\left. - \frac{x^2}{a^2 - \mu} - \frac{y^2}{b^2 - \mu} - \frac{z^2}{c^2 - \mu} \right\} = 0.$$

Consequently, the surfaces $\mathscr{L}$, $\mathscr{M}$, and $\mathscr{N}$ are orthogonal.

Thus, elliptic coordinates are an example of orthogonal curvilinear co-ordinates in space.

We find an expression for the arclength differential in elliptic coordinates. Taking the logarithms of (3_1) and (3_2) and then taking the differentials, we get that

$$2\frac{dx}{x} = \frac{d\lambda}{\lambda - a^2} + \frac{d\mu}{\mu - a^2} + \frac{d\nu}{\nu - a^2},$$

$$2\frac{dy}{y} = \frac{d\lambda}{\lambda - b^2} + \frac{d\mu}{\mu - b^2} + \frac{d\nu}{\nu - b^2}, \qquad (4)$$

$$2\frac{dz}{z} = \frac{d\lambda}{\lambda - c^2} + \frac{d\mu}{\mu - c^2} + \frac{d\nu}{\nu - c^2}.$$

These formulas enable us to express $ds^2 = dx^2 + dy^2 + dz^2$ in terms of the differentials of the elliptic coordinates. The final formula has the form

$$ds^2 = L^2\, d\lambda^2 + M^2\, d\mu^2 + N^2\, d\nu^2,$$

since the terms with products of different differentials disappear by virtue of the orthogonality of the coordinate surfaces.

As for the coefficients L, M, and N, they are not difficult to find. For example, (4) can be used to get the following expression for L^2:

$$4L^2 = \frac{x^2}{(a^2 - \lambda)^2} + \frac{y^2}{(b^2 - \lambda)^2} + \frac{z^2}{(c^2 - \lambda)^2}.$$

By (3_1) and (3_2), this gives us that

$$4L^2 = \frac{(a^2 - \mu)(a^2 - \nu)}{(a^2 - \lambda)(a^2 - b^2)(a^2 - c^2)}$$

$$+ \frac{(b^2 - \mu)(b^2 - \nu)}{(b^2 - \lambda)(b^2 - c^2)(b^2 - a^2)}$$

$$+ \frac{(c^2 - \mu)(c^2 - \nu)}{(c^2 - \lambda)(c^2 - a^2)(c^2 - b^2)},$$

which after simplifications takes the form

$$4L^2 = \frac{(\lambda - \mu)(\lambda - \nu)}{(a^2 - \lambda)(b^2 - \lambda)(c^2 - \lambda)}. \tag{5_1}$$

Similarly,

$$4M^2 = \frac{(\mu - \nu)(\mu - \lambda)}{(a^2 - \mu)(b^2 - \mu)(c^2 - \mu)},$$

$$4N^2 = \frac{(\nu - \lambda)(\nu - \mu)}{(a^2 - \nu)(b^2 - \nu)(c^2 - \nu)}. \tag{5_2}$$

The lack of uniqueness in the correspondence between the elliptic and Cartesian coordinates can be removed. With this goal we express the coordinates λ, μ, ν in terms of elliptic functions.

We let

$$-s = \wp(U) + A \tag{6}$$

and determine the invariants g_2 and g_3 of the function $\wp$ from the condition that

$$4(a^2 - s)(b^2 - s)(c^2 - s) = 4\wp^3(U) - g_2\wp(U) - g_3. \tag{7}$$

Replacing s by a^2, b^2, and c^2 in (6) and, correspondingly, $\wp(U)$ by e_3, e_2, and e_1, we get that

$$-a^2 = e_3 + A, \qquad -b^2 = e_2 + A, \qquad -c^2 = e_1 + A,$$

and so $A = -\frac{1}{3}(a^2 + b^2 + c^2)$. Consequently,

$$e_3 = \frac{1}{3}(a^2 + b^2 + c^2) - a^2 = \frac{b^2 + c^2 - 2a^2}{3},$$

$$e_2 = \frac{1}{3}(a^2 + b^2 + c^2) - b^2 = \frac{c^2 + a^2 - 2b^2}{3},$$

$$e_1 = \frac{1}{3}(a^2 + b^2 + c^2) - c^2 = \frac{a^2 + b^2 - 2c^2}{3}.$$

We have the real case here, since e_1, e_2, and e_3 are real. Therefore, ω is a positive number, and ω' an imaginary number. In place of the elliptic coordinates λ, μ, ν we introduce the parameters u, v, w, replacing s by λ, μ, and ν in (6), and U by u, v, and w, respectively. We get

$$\lambda = \frac{a^2 + b^2 + c^2}{3} - \wp(u),$$

$$\mu = \frac{a^2 + b^2 + c^2}{3} - \wp(v),$$

$$\nu = \frac{a^2 + b^2 + c^2}{3} - \wp(w).$$

We now substitute these expressions in (3_1) and (3_2) for x^2, y^2, and z^2. This gives us

$$x^2 = \frac{[\wp(u) - e_3][\wp(v) - e_3][\wp(w) - e_3]}{(e_2 - e_3)(e_1 - e_3)},$$

$$y^2 = \frac{[\wp(u) - e_2][\wp(v) - e_2][\wp(w) - e_2]}{(e_1 - e_2)(e_3 - e_2)},$$

$$z^2 = \frac{[\wp(u) - e_1][\wp(v) - e_1][\wp(w) - e_1]}{(e_2 - e_1)(e_3 - e_1)}.$$

As we know (see §15), the right-hand sides are the squares of certain meromorphic functions.

Therefore, taking square roots, we obtain x, y, and z as certain single-valued functions of the parameters u, v, and w. Taking the $+$ sign in front of the square roots, we arrive at the following formulas:

$$x = e^{-\eta_3 \omega_3} \sigma^2(\omega_3) \frac{\sigma_3(u)\sigma_3(v)\sigma_3(w)}{\sigma(u)\sigma(v)\sigma(w)},$$

$$y = e^{-\eta_2 \omega_2} \sigma^2(\omega_2) \frac{\sigma_2(u)\sigma_2(v)\sigma_2(w)}{\sigma(u)\sigma(v)\sigma(w)},$$

$$z = e^{-\eta_1 \omega_1} \sigma^2(\omega_1) \frac{\sigma_1(u)\sigma_1(v)\sigma_1(w)}{\sigma(u)\sigma(v)\sigma(w)}.$$

Let us replace u by $u + 2\omega_1$ in these formulas without changing v and w. Since

$$\frac{\sigma_3(u + 2\omega_1)}{\sigma(u + 2\omega_1)} = -\frac{\sigma_3(u)}{\sigma(u)},$$

$$\frac{\sigma_2(u + 2\omega_1)}{\sigma(u + 2\omega_1)} = -\frac{\sigma_2(u)}{\sigma(u)},$$

$$\frac{\sigma_1(u + 2\omega_1)}{\sigma(u + 2\omega_1)} = +\frac{\sigma_1(u)}{\sigma(u)},$$

this substitution does not change z, and x and y only change signs. The quantities x, y, and z undergo analogous changes when other periods are added to the parameter u, as shown by the table:

u	$u + 2\omega_1$	$u + 2\omega_2$	$u + 2\omega_3$
x	$-x$	$-x$	x
y	$-y$	y	$-y$
z	z	$-z$	$-z$

On the other hand, all three coordinates x, y, and z change their signs when u changes sign.

Thus, the adopted representation of the Cartesian coordinates in terms of the parameters u, v, and w works for arbitrary combinations of signs of these Cartesian coordinates.

§58. The Laplace equation in elliptic coordinates

If in the Laplace equation

$$\frac{\partial^2 \varphi}{\partial x^2} + \frac{\partial^2 \varphi}{\partial y^2} + \frac{\partial^2 \varphi}{\partial z^2} = 0$$

orthogonal curvilinear coordinates λ, μ, ν are introduced instead of the Cartesian coordinates x, y, z, then the transformed equation has the form

$$\frac{\partial}{\partial \lambda}\left(\frac{MN}{L}\frac{\partial \varphi}{\partial \lambda}\right) + \frac{\partial}{\partial \mu}\left(\frac{NL}{M}\frac{\partial \varphi}{\partial \mu}\right) + \frac{\partial}{\partial \nu}\left(\frac{LM}{N}\frac{\partial \varphi}{\partial \nu}\right) = 0,$$

where L, M, and N are the functions appearing in the expression of the arclength differential in these coordinates:

$$ds^2 = L^2\,d\lambda^2 + M^2\,d\mu^2 + N^2\,d\nu^2.$$

This fact, which goes back to Lamé (1834), is usually proved in courses in vector analysis, and we may assume it is known.

Let us take λ, μ, ν to be elliptic coordinates and use the expressions found in §55 for L, M, and N. We then get the equation

$$(\mu - \nu)R(\lambda)\frac{\partial}{\partial \lambda}\left\{R(\lambda)\frac{\partial \varphi}{\partial \lambda}\right\}$$

$$+ (\nu - \lambda)R(\mu)\frac{\partial}{\partial \mu}\left\{R(\mu)\frac{\partial \varphi}{\partial \mu}\right\}$$

$$+ (\lambda - \mu)R(\nu)\frac{\partial}{\partial \nu}\left\{R(\nu)\frac{\partial \varphi}{\partial \nu}\right\} = 0,$$

where

$$R(\rho) = \sqrt{(a^2 - \rho)(b^2 - \rho)(c^2 - \rho)}.$$

We introduce instead of λ, μ, ν the quantities u, v, w. The transition functions were given in §57. By virtue of these formulas our equation becomes

$$\{\wp(v) - \wp(w)\}\frac{\partial^2 \varphi}{\partial u^2} + \{\wp(w) - \wp(u)\}\frac{\partial^2 \varphi}{\partial v^2}$$

$$+ \{\wp(u) - \wp(v)\}\frac{\partial^2 \varphi}{\partial w^2} = 0, \tag{1}$$

We look for a particular solution of this equation in the form

$$\varphi = U(u)V(v)W(w). \tag{2}$$

Substituting this in (1), we get

$$U''VW\{\wp(v) - \wp(w)\} + V''WU\{\wp(w) - \wp(u)\}$$
$$+ W''UV\{\wp(u) - \wp(v)\} = 0,$$

from which

$$\frac{U''}{U} = \frac{V''}{V}\frac{\wp(w) - \wp(u)}{\wp(w) - \wp(v)} + \frac{W''}{W}\frac{\wp(u) - \wp(v)}{\wp(w) - \wp(v)}. \tag{3'}$$

The right-hand side is an entire linear function of $\wp(u)$. Consequently,

$$U''/U = A + B\wp(u), \tag{3''}$$

where A and B are constants. Moreover, a comparison of the right-hand sides of (3') and (3'') gives us that

$$A = \frac{V''}{V}\frac{\wp(w)}{\wp(w) - \wp(v)} - \frac{W''}{W}\frac{\wp(v)}{\wp(w) - \wp(v)},$$
$$B = \frac{W''}{W}\frac{1}{\wp(w) - \wp(v)} - \frac{V''}{V}\frac{1}{\wp(w) - \wp(v)}.$$

Multiplying the second of these equations by $\wp(v)$ and adding it to the first, we get that

$$A + B\wp(v) = V''/V$$

and, analogously,

$$A + B\wp(w) = W''/W.$$

We have thus arrived at the following result: the function (2) is a particular integral of equation (1) if the functions U, V, and W satisfy the equation

$$d^2y/dx^2 = [A + B\wp(x)]y \tag{4}$$

for $x = u$, v, and w, respectively, and A and B are constants.

Equation (4) was first studied by Lamé. He showed that this equation can be integrated in elliptic functions for suitably chosen A and for $B = n(n + 1)$, where n is a positive integer. For each positive integer n he found a particular solution of the Laplace equation, and then he used these particular solutions to form the general solution.

Further investigations of equation (4) are due to Hermite. Some of his results are presented in §59.

§59. The Lamé equation

This equation is a special case of a homogeneous linear differential equation whose coefficients are elliptic functions with the same periods. For such equations there is a general theorem due to Picard. This theorem asserts that *if the general integral of such an equation is a meromorphic function, then this equation can be integrated with the help of doubly periodic functions of the second kind with the same periods.* Furthermore, following Hermite, one calls a (meromorphic) function $f(u)$ a *doubly periodic function of the second kind* with periods 2ω and $2\omega'$ if

$$f(u + 2\omega) = \mu f(u), \qquad f(u + 2\omega') = \mu' f(u),$$

where μ and μ' are constants (the so-called *multipliers* of our function).

Since we are interested in the Lamé equation, we prove the Picard theorem for the case of a second-order equation. The reader can easily see how to carry the proof over to the general case.

Accordingly, suppose that we are given the equation

$$y'' + f(x)y' + g(x)y = 0, \tag{1}$$

where $f(x)$ and $g(x)$ are elliptic functions with periods 2ω and $2\omega'$, and suppose that the general integral of this equation is a meromorphic function of x.

We take a particular integral $y = \varphi(x)$. The functions $\varphi(x + 2\omega)$ and $\varphi(x + 4\omega)$ are also particular integrals of our equation. But since the equation has only two linearly independent particular integrals, it follows that

$$\varphi(x + 4\omega) = \alpha\varphi(x) + \beta\varphi(x + 2\omega), \tag{2}$$

where α and β are certain (definite) constants.

We now consider the function

$$\psi(x) = \lambda_1 \varphi(x) + \lambda_2 \varphi(x + 2\omega),$$

which is a particular integral of (1) for arbitrary constants λ_1 and λ_2. In view of (2),

$$\psi(x + 2\omega) = \lambda_1 \varphi(x + 2\omega) + \lambda_2[\alpha\varphi(x) + \beta\varphi(x + 2\omega)]$$
$$= \alpha\lambda_2\varphi(x) + (\lambda_1 + \beta\lambda_2)\varphi(x + 2\omega).$$

We require that $\psi(x + 2\omega) = \mu\psi(x)$, where μ is a constant. This requirement leads to the following relations:

$$\alpha\lambda_2 = \mu\lambda_1, \qquad \beta\lambda_2 + \lambda_1 = \mu\lambda_2. \tag{3}$$

This yields an equation for determining μ:

$$\mu^2 - \mu\beta - \alpha = 0.$$

When μ has been found, system (3) enables us to find the ratio λ_1/λ_2: it is equal to $\mu - \beta$. Thus we have found a particular integral $\psi(x)$ for which

$$\psi(x + 2\omega) = \mu\psi(x). \tag{4}$$

Starting from this particular integral and dealing with it in the same way as we dealt with $\varphi(x)$ but taking $2\omega'$ instead of 2ω, we find a combination

$$F(x) = \lambda_1'\psi(x) + \lambda_2'\psi(x + 2\omega')$$

such that $F(x + 2\omega') = \mu'F(x)$, where μ' is again some constant. Since, by (4),

$$F(x + 2\omega) = \mu F(x),$$

it follows that $F(x)$ is a doubly periodic function of the second kind. But this is a particular integral of our equation.

With the help of this integral we lower the order of the equation. For this let $y = F(x) \int z \, dx$. Our equation becomes

$$F(x)z' + 2F'(x)z + f(x)F(x)z = 0,$$

or

$$z' + \left[2\frac{F'(x)}{F(x)} + f(x) \right] z = 0.$$

The coefficients of this equation are elliptic functions with the previous periods.

If we had considered an equation of order n ($n \neq 2$), then we would have arrived at an equation of order $n - 1$ and we would have had to repeat for it the arguments that were applied to the original equation. We would have gotten that the new equation has at least one particular integral that is a doubly periodic function of the second kind.

However, in the present case z can be found directly:

$$z = \frac{1}{[F(x)]^2} e^{-\int f(x)dx},$$

and it is easy to see that this is a doubly periodic function of the second kind.

Thus, Picard's theorem is proved for this special case.

Let us now turn to the Lamé equation

$$d^2y/du^2 = [n(n+1)\wp(u) + l]y. \tag{5}$$

We assume that n is a positive integer, but we do not impose any restrictions on the constant l.

The general integral of this equation is a meromorphic function. For a proof, observe that only the poles of $\wp(u)$ can be singular points of this integral at a finite distance.

For example, let us take the pole $u = 0$ and, following general methods in the theory of differential equations, find particular integrals of our equation in the form of series in powers of u. We construct series that formally satisfy our equation. These series will contain only integer powers of u and only finitely many terms with negative powers.

Since it can be proved that the series converge in some neighborhood of the point $u = 0$ (with the possible exception of the point $u = 0$ itself), the point $u = 0$ is either a regular point or a pole for the integral. Applying the same argument to the other poles of $\wp(u)$, we prove that the general integral of our equation is meromorphic.

Accordingly, we try to find series that formally satisfy the Lamé equation.

For convenience let

$$n(n+1)\wp(u) + l = \frac{n(n+1)}{u^2} + \sum_{r=1}^{\infty} A_{2r}u^{2r-2}$$

and $y = u^{\alpha} \sum_0^{\infty} C_k u^k$ ($C_0 \neq 0$). Substituting these series in the equation, we get that

$$\sum_{k=0}^{\infty} C_k(k+\alpha)(k+\alpha-1)u^{k+\alpha-2}$$

$$= \left[n(n+1) + \sum_{r=1}^{\infty} A_{2r}u^{2r} \right] \sum_{k=0}^{\infty} C_k u^{k+\alpha-2}.$$

A comparison of the coefficients gives us that

$$C_0\omega(\alpha) = 0,$$
$$C_1\omega(\alpha + 1) = 0,$$
$$C_2\omega(\alpha + 2) + C_0A_2 = 0,$$
$$C_3\omega(\alpha_3) + C_1A_2 = 0,$$

$$\cdots\cdots\cdots\cdots\cdots\cdots\cdots$$

$$C_{2m}\omega(\alpha + 2m) + C_{2m-2}A_2 + C_{2m-4}A_4 + \cdots + C_0A_{2m} = 0,$$
$$C_{2m-1}\omega(\alpha + 2m - 1) + C_{2m-3}A_2 + C_{2m-5}A_4 + \cdots + C_1A_{2m-2} = 0,$$

where $\omega(x) = x(x - 1) - n(n + 1)$. Since $C_0 \neq 0$, α must be a root of $\omega(x)$, from which either $\alpha = n + 1$ or $\alpha = -n$.

Under each of these assumptions we can take $C_1 = C_3 = C_5 = \cdots = 0$. To determine the coefficients C_{2k} we have the equation

$$C_{2m}\omega(\alpha + 2m) + C_{2m-2}A_2 + \cdots + C_0A_{2m} = 0,$$

which enables us to find all the C_{2k} if $\omega(\alpha + 2k)$ never vanishes. But, as we have seen, the equation $\omega(x) = 0$ has the roots $x = n + 1$ and $x = -n$, whose difference is an odd number. And since $\omega(\alpha) = 0$, it follows that $\omega(\alpha + 2k)$ is nonzero for an arbitrary positive integer k. We see that the series contains only integer power of the variable, and the number of terms with negative powers is finite. The assertion can thus be regarded as proved.

We use Picard's theorem.

Let $\varphi(u)$ be a particular integral of (5) that is a doubly periodic function of the second kind with multipliers μ and μ'. We introduce the function

$$\varphi(u)e^{-\lambda u}\frac{\sigma(u)}{\sigma(u - a)} = \psi(u),$$

where λ and a are undetermined constants. Adding the periods 2ω and $2\omega'$ to u, we get that

$$\psi(u + 2\omega) = \mu\varphi(u)e^{-\lambda(u+2\omega)}e^{2\eta a}\frac{\sigma(u)}{\sigma(u - a)} = \mu e^{2\eta a}e^{-2\lambda\omega}\psi(u)$$

and

$$\psi(u + 2\omega') = \mu'e^{2\eta' a}e^{-2\lambda\omega'}\psi(u).$$

We now determine a and λ in such a way that $\psi(u)$ has periods 2ω and $2\omega'$. For this it is necessary that

$$e^{-2(a\eta-\lambda\omega)} = \mu, \qquad e^{-2(a\eta'-\lambda\omega')} = \mu',$$

or

$$\omega\lambda - \eta a = \tfrac{1}{2}\ln\mu, \qquad \omega'\lambda - \eta'a = \tfrac{1}{2}\ln\mu'.$$

The determinant $\eta'\omega - \eta\omega'$ of this system is nonzero. Hence, the required quantities a and λ exist.

Since $\psi(u)$ is an elliptic function, it can be expressed in terms of the sigma function from the poles and zeros.

We assume that $\varphi(u)$ has an n-fold pole at $u = 0$. Therefore, $\psi(u)$ has a pole of order $n - 1$ at this point. Moreover, $\psi(u)$ has a simple pole at $u = a$. Let $a_1, \ldots, a_n$ be the complete system of zeros of $\varphi(u)$, and let $a_1 + \cdots + a_n = a$; then

$$\psi(u) = C \frac{\sigma(u - a_1)\sigma(u - a_2) \cdots \sigma(u - a_n)}{\sigma(u - a)[\sigma(u)]^{n-1}}.$$

Consequently,

$$\varphi(u) = C e^{\lambda u} \frac{\sigma(u - a_1)\sigma(u - a_2) \cdots \sigma(u - a_n)}{[\sigma(u)]^n}.$$

The formula obtained contains the constants λ and $a_1, \ldots, a_n$, which must be determined in such a way that the function satisfies (5).

This determination of the constants is especially simple for $n = 1$. We dwell on this case in detail.

The Lamé equation is

$$y'' = [2\wp(u) + l]y. \tag{6}$$

Here the particular integral has the form

$$y_1 = e^{-\lambda u} \frac{\sigma(u + a)}{\sigma(u)}.$$

Desiring to substitute this function in the Lamé equation (6), we find the logarithmic derivative of y_1:

$$y_1'/y_1 = -\lambda + \zeta(u + a)\zeta(u);$$

further, we find that

$$y_1''/y_1 = \wp(u) - \wp(u + a) + [-\lambda + \zeta(u + a) - \zeta(u)]^2.$$

The identity

$$2\wp(u) + l = \wp(u) - \wp(u + a) + [-\lambda + \zeta(u + a) - \zeta(u)]^2$$

follows from (6). It can be represented in the form

$$\wp(u) + \wp(u + a) + l = [-\lambda + \zeta(u + a) - \zeta(u)]^2,$$

or

$$l - \wp(a) + \frac{1}{4}\left[\frac{\wp'(u) - \wp'(a)}{\wp(u) - \wp(a)}\right]^2$$

$$= \left[\zeta(a) - \lambda + \frac{1}{2}\frac{\wp'(u) - \wp'(a)}{\wp(u) - \wp(a)}\right]^2.$$

This identity holds if and only if $\wp(a) = l$ and $\zeta(a) = \lambda$. Thus, for $n = 1$ the Lamé equation has the particular integral

$$y_1 = e^{-u\zeta(a)}\sigma(u + a)/\sigma(u) \quad (\wp(a) = l).$$

Replacing a by $-a$, we find another particular integral:

$$y_2 = e^{u\zeta(a)}\sigma(u - a)/\sigma(u).$$

They are linearly independent if a is not a half-period, i.e., if $l \neq e_k$ $(k = 1, 2, 3)$.

§60. The Picard theorems on meromorphic functions

In 1879 Picard proved two theorems that turned out to have an essential influence on the subsequent development of the theory of functions of a complex variable.

The first theorem of Picard is as follows:

If a function $f(z)$ meromorphic in the open plane omits (i.e., does not take on) more than two values, then it is a constant.

We note that examples of meromorphic functions omitting two values are very easy to construct with the help of the function e^z, which omits the values 0 and ∞. For example, the function

$$(a - be^z)/(1 - e^z) \quad (a \neq b)$$

does not take the two values a and b.

We give a proof of Picard's theorem due to Picard himself.

Suppose that the meromorphic function $f(z)$ omits the three values a, b, and c (one of which can be infinite). In this case

$$g(z) = \frac{c - b}{c - a} \cdot \frac{f(z) - a}{f(z) - b}$$

omits the values 0, 1, and ∞, and hence is an entire function omitting the values 0 and 1.

We recall the modular function $\lambda = \lambda(\tau)$ considered in §23. This function maps the domain D_2 (see Figure 8 in Chapter IV) onto the plane, cut along the semiaxes $(1, \infty)$ and $(-\infty, 0)$.

The inverse function $\tau = \tau(\lambda)$, considered in the uncut λ-plane, takes infinite values, and its only critical points are $\lambda = 0$, 1, and ∞. Upon a circuit of each of these points the quantity τ is subjected to some transformations of the group Σ_2.

Let us now consider

$$\tau[g(z)] = G(z) \tag{1}$$

as a function of z. Suppose that the point z describes a continuous curve, starting from the initial position z_0. If we choose one of the possible values of $G(z)$ at z_0, and then define it with continuity observed, then we get first of all a function that is analytic in a neighborhood of z_0. This extension process can be implemented along an arbitrary path from z_0 to an arbitrary point z_1, since the function $g(z)$ never takes one of the critical values 0, 1, and ∞, and hence the function $\tau(\lambda)$ is analytic at all points of the curve described by the point

$\lambda = g(z)$. Since the whole plane is a simply connected domain, the monodromy theorem tells us that $G(z)$ is a single-valued function, and hence is itself an entire function.[45]

Since $\Im \tau > 0$, the function (1) satisfies the inequality $\Im G(z) > 0$ everywhere in the z-plane. Therefore, the function

$$e^{iG(z)} \tag{2}$$

satisfies the inequality $|e^{iG(z)}| < 1$ in the whole plane of the variable z.

By Liouville's theorem, the function (2) must be a constant, hence $g(z)$ is a constant, and hence the original function $f(z)$ is a constant.

The second theorem, which later acquired the name of the big Picard theorem, is formulated as follows:

If z_0 is an essential singular point of a single-valued function $f(z)$ and is isolated from other essential singular points of this function, then $f(z)$ omits at most two values in an arbitrary neighborhood of z_0.

The proof of the big Picard theorem is technically more complicated, but is conceptually analogous to the proof of his first theorem, and we need not present it.

The reader has possibly been struck by the discrepancy between the simplicity of the statement of Picard's theorem and its proof, which uses a very special function. This circumstance was of course noted at once, and it stimulated the development of new methods.

The first "elementary" proof of the Picard theorem was given by Borel in 1896.

At present the circle of questions associated with the Picard theorems represents one of the most interesting and beautiful parts of function theory, a part rich in deep results and remarkable general methods.

It should be mentioned that the existence of various estimates is established in certain propositions relating to this circle of ideas. Here it turns out that the introduction of modular functions is unavoidable if one wishes to obtain sharp estimates. One of the most remarkable theorems of this kind is presented in the next section.

§61. The Landau theorem

This theorem can be formulated as follows: *If the power series*

$$f(z) = a_0 + a_n z^n + a_{n+1} z^{n+1} + \cdots \qquad (a_n \neq 0, \ n \geq 1)$$

---

[45] It is not hard to prove directly that $G(z)$ is a single-valued. Indeed, if we obtained different values for $G(z_1)$ on two different paths from z_0 to z_1, so that $G(z)$ would not return to the original value when z moves along the closed contour formed by these paths, then this would mean that in its plane the point $\lambda = g(z)$ describes a closed contour with at least one of the two critical points $\lambda = 0$ or $\lambda = 1$ contained inside it. However, in this case the path from z_0 to z_1 could be deformed in such a way (here we use the fact that the plane is simply connected) that it would pass through at least one of these critical points; but this is excluded, because $g(z)$ does not take the values 0 and 1.

*converges in the disk of radius R about the point $z = 0$ and does not take
the values 0 and 1, then the radius R of this disk does not exceed a certain
quantity depending only on a_0, a_n, and n:*

$$R \le \varphi(a_0, a_n, n).$$

As we shall see, this theorem establishes a certain estimate given by the
function $\varphi(a_0, a_n, n)$. In the process of proving the Landau theorem we
find an exact expression for this function.[46]

The connection of the Landau theorem with the Picard theorem in §60
is obvious.

Indeed, if an entire function $g(z)$ not taking the values 0 and 1 is not a
constant, then the series of coefficients $a_0, a_1, \ldots$, beginning with the second coefficient, contains a first nonzero coefficient, say a_n. By the Landau
theorem there is a disk of finite radius on whose circle $g(z)$ has at least
one singular point, and this contradicts an assumption.

Thus, the Picard theorem is a consequence of the Landau theorem.

We proceed to a proof of the Landau theorem. Let[47]

$$\Omega(z) = \frac{\tau[f(z)] - \tau(a_0)}{\tau[f(z)] - \overline{\tau(a_0)}}.$$

Here $\tau(a_0)$ is one of the possible values of the function $\tau[f(z)]$ at $z = 0$. The values of $\tau[f(z)]$
at other points of the disk $|z| < R$ are chosen in such a way that continuity is observed. The
function $\Omega(z)$ is regular in the domain $|z| < R$, a fact proved just like the corresponding
statement in the Picard theorem. Further, as is easy to see, $|\Omega(z)| < 1$ for $|z| < R$, and
$\Omega(0) = 0$. Since $f(z) = a_0 + a_n z^n + \cdots$, $\Omega(z)$ has an expansion of the form

$$\Omega(z) = \frac{\tau'(a_0)}{2i\Im\tau(a_0)} a_n z^n + \cdots.$$

In view of the Cauchy inequalities we have

$$\left| \frac{\tau'(a_0)}{2\Im\tau(a_0)} a_n \right| \le \frac{1}{R^n},$$

and hence

$$R \le \sqrt[n]{\frac{2\Im\tau(a_0)}{|a_n| \cdot |\tau'(a_0)|}};$$

it remains to prove that the estimate is sharp, i.e., that

$$\varphi(a_0, a_n, n) = \sqrt[n]{\frac{2\Im\tau(a_0)}{|a_n| \cdot \tau'(a_0)|}}.$$

If we want the equality

$$R = \sqrt[n]{\frac{2\Im\tau(a_0)}{|a_n| \cdot |\tau'(a_0)|}}$$

---

[46]Landau proved his theorem in 1904. The proof we present, which leads to an exact
expression for $\varphi(a_0, a_n, n)$, is due to Carathéodory (1905).

[47]The function $\tau(\lambda)$ was introduced in §60.

to hold for some function $f(z)$, then it is necessary that for the corresponding function $\Omega(z) = A_n z^n + \cdots$, which satisfies the inequality $|\Omega(z)| < 1$ for $|z| < R$, we have the equality $|A_n| = 1/R^n$. But this means that the modulus of the function

$$\Omega(z)/z^n, \qquad\qquad (\alpha)$$

which is regular in the disk, must attain its maximum at $z = 0$. Consequently, the function (α) must be a constant:

$$\Omega(z) = \frac{\varepsilon z^n}{R^n} \qquad (|\varepsilon| = 1).$$

From the equality

$$\frac{\varepsilon z^n}{R^n} = \frac{\tau[f(z)] - \tau(a_0)}{\tau[f(z)] - \overline{\tau(a_0)}}$$

we find first that

$$\tau[f(z)] = \frac{\tau(a_0)R^n - \overline{\tau(a_0)}\varepsilon z^n}{R^n - \varepsilon z^n},$$

and then that

$$f(z) = \lambda\left(\frac{\tau(a_0)R^n - \overline{\tau(a_0)}\varepsilon z^n}{R^n - \varepsilon z^n}\right).$$

It has thus been proved that the quality

$$R = \sqrt[n]{\frac{2\Im\tau(a_0)}{|a_n| \cdot |\tau'(a_0)|}}$$

is possible, i.e., the estimate is sharp.

§62. On meromorphic functions having an algebraic addition theorem

A meromorphic function $\varphi(z)$ has an algebraic addition theorem if there is an identity

$$F(\varphi(z_1 + z_2), \varphi(z_1), \varphi(z_2)) = 0, \qquad\qquad (1)$$

where F is an entire rational function of its arguments.

Above we saw that every elliptic function has an algebraic addition theorem. Algebraic addition theorems also hold (as can easily be verified directly) for degeneracies of elliptic functions, examples of which are rational functions of z and rational functions of $e^{\pi i z/\omega}$.

Weierstrass proved that there are no other functions having an algebraic addition theorem.

We present a proof, due to Osgood, of this theorem of Weierstrass.

Suppose that the meromorphic function $\varphi(z)$ has an algebraic addition theorem (1), where the polynomial F has degree m in the first argument. Moreover, suppose that $\varphi(z)$ is not a rational function, and hence $z = \infty$ is an essential singularity of it.

Take a number C different from the values omitted by $\varphi(z)$ in a neighborhood of $z = \infty$. In this case (here the big Picard theorem is used) there are $m + 1$ different points a_k such that $\varphi(a_k) = C$ $(k = 0, 1, \ldots, m)$. We then choose a regular point ζ of $\varphi(z)$ such that the points $a_k + \zeta$ $(k = 0, 1, \ldots, m)$ are also regular for $\varphi(z)$.

Obviously, the same holds for all z sufficiently close to ζ. Taking z to be one of the indicated values, we consider the equation

$$F(x, \varphi(z), C) = 0. \tag{2}$$

This equation has $m + 1$ roots $x = \varphi(z + a_k)$. In view of our condition these roots cannot all be distinct. Hence, for each z in some neighborhood of ζ we have the equality

$$\varphi(z + a_r) = \varphi(z + a_s).$$

The pair (a_r, a_s) can change in passing from one value of z to another. But since there are infinitely many of these values z, while there are finitely many of the pairs (a_r, a_s), for at least one pair (a_ρ, a_σ) the equality

$$\varphi(z + a_\rho) = \varphi(z + a_\sigma) \tag{3}$$

holds for infinitely many points z, and these points z have a limit point ζ at which $\varphi(z)$ is regular. In view of the uniqueness theorem for analytic functions, this implies that (3) is an identity. But this means that $\varphi(z)$ has period $a_\sigma - a_\rho$.

Thus, $\varphi(z)$ is periodic. If $\varphi(z)$ is doubly periodic, the theorem is proved. Therefore, it remains to consider the second possible case, when $\varphi(z)$ is simply periodic. Assuming for simplicity that its primitive period is 2π, we must prove that in this case $\varphi(z)$ is a rational function of $w = e^{iz}$. The mapping $w = e^{iz}$ carries the strip $0 < \Re z < 2\pi$ into the w-plane, cut along the semi-axis $(0, \infty)$, and it is not hard to see that $\varphi(z) = \psi(w)$ is meromorphic in the w-plane, punctured at the points $w = 0, \infty$. If we assume that at least one of these points is an essential singularity for $\psi(w)$, then the Picard theorem can again be applied to $\psi(w)$ and, as above, $m + 1$ *distinct* points b_k at which $\psi(w) = C$ can be found. The inverse images of these points lie in the strip $0 \leq \Re z < 2\pi$. Therefore, the treatment of equation (2) presented above shows that $\varphi(z)$ has a period whose real part is $< 2\pi$, which is impossible.

The proof of the Weierstrass theorem is thus complete.

§63. On Fourier series of analytic functions

In this section we consider functions with period 2π that are analytic on the real axis.

Each such function $f(z)$ has some strip of regularity, i.e., the largest strip bounded by two lines $y = -\alpha$ and $y = \beta$ $(\alpha, \beta > 0)$, in whose interior the function is regular. The function $f(z)$ has at least one singular point on each of these lines.

THEOREM 1. *Let $f(z)$ be an analytic function with period 2π that is regular in the closed strip $-a \leq y \leq b$ $(a, b > 0)$. Further, let $\sum_{-\infty}^{\infty} c_k e^{ikx}$ be the Fourier series of $f(x)$. Then*

$$\left.\begin{array}{l} |c_{-k}| \leq Me^{-kb}, \\ |c_k| \leq Me^{-ka} \end{array}\right\} \qquad (k = 0, 1, 2, \ldots), \qquad (1)$$

where M is a constant (independent of k).

PROOF. We consider the rectangle with vertices $(0, 0)$, $(2\pi, 0)$, $(2\pi, b)$ and $(0, b)$. The function $f(z)e^{ikz}$ is regular inside and on the boundary of this rectangle. Therefore, applying Cauchy's theorem to this function, we get the equality

$$\int_0^{2\pi} f(x)e^{ikx}\, dx + i\int_0^b f(2\pi + iy)e^{-ky}\, dy$$

$$+ \int_{2\pi}^0 f(x + ib)e^{ikx - kb}\, dx + i\int_b^0 f(iy)e^{-ky}\, dy = 0.$$

By the periodicity of $f(z)$, the second and fourth integrals cancel each other. Therefore,

$$2\pi c_{-k} = \int_0^{2\pi} f(x)e^{ikx}\, dx = e^{-kb}\int_0^{2\pi} f(x + ib)e^{ikx}\, dx.$$

This yields the inequalities

$$|c_{-k}| \leq Me^{-kb} \qquad (k = 0, 1, 2, \ldots),$$

where M is the maximum of the modulus of $f(z)$ in the strip under consideration.

The proof of the inequalities for c_k is analogous.

THEOREM 2. *If the inequalities*

$$|c_{-k}| \leq Me^{-kb}, \quad |c_k| \leq Me^{-ka} \qquad (a, b > 0)$$

hold, then $\sum_{-\infty}^{\infty} c_k e^{ikz}$ converges uniformly and absolutely in an arbitrary strip lying entirely inside the strip

$$-a < y < b \qquad (2)$$

and represents a regular analytic function with period 2π inside the strip (2).

The proof is so simple that we can omit it.

COROLLARY. *The quantities α and β determining the strip of regularity of $f(z)$ are equal to*

$$\alpha = \varliminf_{k \to \infty} \ln \frac{1}{\sqrt[k]{|c_k|}}, \tag{3_1}$$

$$\beta = \varliminf_{k \to \infty} \ln \frac{1}{\sqrt[k]{|c_{-k}|}}. \tag{3_2}$$

PROOF. Let α and β be defined by (3_1) and (3_2). In this case

$$|c_{-k}| < e^{-k(\beta - \varepsilon)}, \qquad |c_k| < e^{-k(\alpha - \varepsilon)}$$

for an arbitrary $\varepsilon > 0$ and for all sufficiently large k. On the basis of Theorem 2, this implies that $f(z)$ is regular in the strip $-(\alpha - \varepsilon) < y < \beta - \varepsilon$, and since ε is arbitrary, $f(z)$ is regular in the strip $-\alpha < y < \beta$.

It remains to show that this is the largest strip where $f(z)$ is regular. Assuming the contrary, we would get by virtue of Theorem 1 that in the inequalities

$$|c_{-k}| \leq M e^{-kb} \qquad (k = 0, 1, 2, \dots)$$

we can take $b > \beta$, or in the inequalities

$$|c_k| \leq M e^{-ka} \qquad (k = 0, 1, 2, \dots)$$

we can take $a > \alpha$. In the first case we would have

$$\varliminf_{k \to \infty} \ln \frac{1}{\sqrt[k]{|c_k|}} > \beta,$$

and in the second

$$\varliminf_{k \to \infty} \ln \frac{1}{\sqrt[k]{|c_k|}} > \alpha.$$

But these two contradict the assumption.

We now remark that if $f(z)$ is an entire function, then the parameters α and β are both infinitely large, and the inequalities (1) hold for arbitrary $a > 0$ and $b > 0$. However, recall that M is a function of a and b that increases without bound as $a \to \infty$ and $b \to \infty$, provided, of course, that $f(z)$ is not a constant.

The theta functions introduced at the very beginning of the book (§3) are excellent examples of entire functions with a real period (such a period can always be reduced to 2π).

We now consider certain meromorphic functions with a real period. Let us take the function

$$\operatorname{sn} \frac{2Kv}{\pi} = \operatorname{sn}\left(\frac{2Kv}{\pi}; k\right),$$

where K is a complete elliptic integral of the first kind for the modulus k. This function has periods 2π and $\pi i K'/K$ and admits an expansion

$$\operatorname{sn}\frac{2Kv}{\pi} = \sum_{n=1}^{\infty} c_n \sin nv.$$

Replacing v by $\pi - v$, we get that

$$\operatorname{sn}\frac{2Kv}{\pi} = \sum_{n=1}^{\infty} (-1)^{n-1} c_n \sin nv;$$

Consequently, $c_{2n} = 0$ $(n = 1, 2, \dots)$, and

$$\operatorname{sn}\frac{2Kv}{\pi} = \sum_{n=1}^{\infty} c_{2n-1} \sin(2n-1)v.$$

To determine the Fourier coefficients c_{2n-1} we have the formula

$$c_{2n-1} = \frac{2}{\pi} \int_0^{\pi} \operatorname{sn}\frac{2Kv}{\pi} \sin(2n-1)v\, dv$$

$$= \frac{2}{\pi i} \int_{-\pi/2}^{\pi/2} \operatorname{sn}\frac{2Kv}{\pi} e^{(2n-1)iv}\, dv.$$

Take the rectangle with vertices at the points

$$-\frac{\pi}{2},\ \frac{\pi}{2},\ \frac{\pi}{2} + \frac{\pi i K'}{K},\ -\frac{\pi}{2} + \frac{\pi i K'}{K}.$$

Integrating the function

$$\operatorname{sn}\frac{2Kv}{\pi} e^{(2n-1)iv} \tag{4}$$

over the boundary of this rectangle and taking into account that the function (4) has period π, by virtue of which the integrals over the lateral sides cancel each other, we find that

$$\frac{2}{\pi i} \int_{-\pi/2}^{\pi/2} \operatorname{sn}\frac{2Kv}{\pi} e^{(2n-1)iv}\, dv$$

$$+ \frac{2}{\pi i} \int_{\pi/2+\pi i K'/K}^{-\pi/2+\pi i K'/K} \operatorname{sn}\frac{2Kv}{\pi} e^{(2n-1)iv}\, dv$$

$$= 4 \operatorname*{Res}_{v=\pi i K'/2K} \left\{ \operatorname{sn}\frac{2Kv}{\pi} e^{(2n-1)iv} \right\},$$

or

$$c_{2n-1}(1 - h^{2n-1}) = 2\pi h^{n-1/2}/kK \qquad (h = e^{-\pi K'/K}),$$

because

$$\operatorname*{Res}_{v=\pi i K'/K} \left\{ \operatorname{sn}\frac{2Kv}{\pi} e^{(2n-1)iv} \right\} = \frac{\pi}{2kK} e^{-(2n-1)\pi K'/2K}.$$

Thus,

$$c_{2n-1} = \frac{2\pi}{kK} \frac{h^{n-1/2}}{1 - h^{2n-1}}.$$

Consequently, our expansion has the form

$$\operatorname{sn} \frac{2Kv}{\pi} = \frac{2\pi}{kK} \sum_{n=1}^{\infty} \frac{h^{n-1/2}}{1 - h^{2n-1}} \sin(2n - 1)v.$$

Here the convergence strip is

$$|\Im v| < \Im \pi i K'/2K.$$

The following expansions are valid in the same strip:

$$\operatorname{cn} \frac{2Kv}{\pi} = \frac{2\pi}{kK} \sum_{n=1}^{\infty} \frac{h^{n-1/2} \cos(2n - 1)v}{1 + h^{2n-1}},$$

$$\operatorname{dn} \frac{2Kv}{\pi} = \frac{\pi}{2K} + \frac{2\pi}{K} \sum_{n=1}^{\infty} \frac{h^{n} \cos 2nv}{1 + h^{2n}}.$$

All three functions have poles on the boundaries of this strip.

Tables of the Most Important Formulas

I. Basic trigonometric functions

$$\sin u = u \prod{}' \left(1 - \frac{u}{m\pi}\right) e^{\frac{u}{m\pi}}$$

$$\operatorname{ctg} u = \frac{1}{u} + \sum{}' \left(\frac{1}{u - m\pi} + \frac{1}{m\pi}\right) = \frac{d}{du}\ln\sin u$$

$$\frac{1}{\sin^2 u} = \frac{1}{u^2} + \sum{}' \frac{1}{(u - m\pi)^2} = -\frac{d}{du}\operatorname{ctg} u$$

II. The Weierstrass functions

$$\sigma(u) = u \prod{}' \left(1 - \frac{u}{s}\right) e^{\frac{u}{s} + \frac{u^2}{2s^2}}$$

$$\zeta(u) = \frac{1}{u} + \sum{}' \left(\frac{1}{u - s} + \frac{1}{s} + \frac{u}{s^2}\right) = \frac{d}{du}\ln\sigma(u)$$

$$\wp(u) = \frac{1}{u^2} + \sum{}' \left\{\frac{1}{(u - s)^2} - \frac{1}{s^2}\right\} = -\zeta'(u)$$

$$g_2 = 60 \sum{}' \frac{1}{s^4} \;;\; g_3 = 140 \sum{}' \frac{1}{s^6}$$

$$(s = 2m\omega + 2m'\omega')$$

$$\sigma(-u) = -\sigma(u); \quad \zeta(-u) = -\zeta(u); \quad \wp(-u) = \wp(u)$$

$$\sigma(u) = u - \frac{g_2 u^5}{2^4 \cdot 3 \cdot 5} - \frac{g_3 u^7}{2^3 \cdot 3 \cdot 5 \cdot 7} - \cdots$$

$$\zeta(u) = \frac{1}{u} - \frac{g_2 u^3}{2^2 \cdot 3 \cdot 5} - \frac{g_3 u^5}{2^2 \cdot 5 \cdot 7} - \cdots$$

$$\wp(u) = \frac{1}{u^2} + \frac{g_2 u^2}{2^2 \cdot 5} + \frac{g_3 u^4}{2^2 \cdot 7} + \cdots$$

III. Homogeneity relations

$$\sigma(u) = \sigma(u \mid \omega, \omega') = \sigma(u; g_2, g_3)$$
$$\zeta(u) = \zeta(u \mid \omega, \omega') = \zeta(u; g_2, g_3)$$
$$\wp(u) = \wp(u \mid \omega, \omega') = \wp(u; g_2, g_3)$$
$$g_2 = g_2(\omega, \omega'); \quad g_3 = g_3(\omega, \omega')$$

$$\sigma(\lambda u \mid \lambda\omega, \lambda\omega') = \lambda\sigma(u \mid \omega, \omega')$$
$$\zeta(\lambda u \mid \lambda\omega, \lambda\omega') = \frac{1}{\lambda}\,\zeta(u \mid \omega, \omega')$$
$$\wp(\lambda u \mid \lambda\omega, \lambda\omega') = \frac{1}{\lambda^2}\,\wp(u \mid \omega, \omega')$$
$$g_2(\lambda\omega, \lambda\omega') = \frac{1}{\lambda^4}\,g_2(\omega, \omega')$$
$$g_3(\lambda\omega, \lambda\omega') = \frac{1}{\lambda^6}\,g_3(\omega, \omega')$$

IV. The differential equation of the function $\wp$

$$\wp'^2(u) = 4\wp^3(u) - g_2\wp(u) - g_3 =$$
$$= 4\{\wp(u) - e_1\}\{\wp(u) - e_2\}\{\wp(u) - e_3\}$$

$$e_1 + e_2 + e_3 = 0$$
$$e_1e_2 + e_2e_3 + e_3e_1 = -\frac{1}{2}\,(e_1^2 + e_2^2 + e_3^2) = -\frac{1}{4}\,g_2$$
$$e_1e_2e_3 = \frac{1}{4}\,g_3$$
$$G = \frac{1}{16}\,(g_2^3 - 27g_3^2) = (e_1 - e_2)^2\,(e_2 - e_3)^2\,(e_3 - e_1)^2$$

$$\omega_1 = \omega, \quad \omega_2 = -\omega - \omega', \quad \omega_3 = \omega'$$
$$e_\alpha = \wp(\omega_\alpha) \quad (\alpha = 1, 2, 3)$$

V. Addition of periods

$$\wp\,(u+2\omega)=\wp\,(u+2\omega')=\wp\,(u)$$

$$\zeta\,(u+2\omega)=\zeta\,(u)+2\eta \qquad \eta=\zeta\,(\omega)=\eta_1$$
$$\zeta\,(u+2\omega')=\zeta\,(u)+2\eta' \qquad \eta'=\zeta\,(\omega')=\eta_3$$

$$\eta\omega'-\eta'\omega=\begin{cases} \dfrac{\pi i}{2} & \left(\Im\,\dfrac{\omega'}{\omega}>0\right) \\[2ex] -\dfrac{\pi i}{2} & \left(\Im\,\dfrac{\omega'}{\omega}<0\right) \end{cases}$$

$$\eta_2=-\eta-\eta'$$

$$\sigma\,(u+2\omega_\alpha)=-e^{2\eta_\alpha\,(u+\omega_\alpha)}\sigma\,(u) \qquad (\alpha=1,\ 2,\ 3)$$

$$\sigma_\alpha\,(u)=-e^{\eta_\alpha u}\,\frac{\sigma\,(u-\omega_\alpha)}{\sigma\,(\omega_\alpha)} \qquad (\alpha=1,\ 2,\ 3)$$

$$\sigma_\alpha\,(u+2\omega_\alpha)=-e^{2\eta_\alpha\,(u+\omega_\alpha)}\sigma_\alpha\,(u) \qquad (\alpha=1,\ 2,\ 3)$$

$$\sigma_\alpha\,(u+2\omega_\beta)=e^{2\eta_\beta\,(u+\omega_\beta)}\sigma_\alpha\,(u) \qquad (\beta\neq\alpha;\ \alpha,\ \beta=1,\ 2,\ 3)$$

$$\zeta_\alpha\,(u)=\frac{d}{du}\ln\sigma_\alpha\,(u) \qquad (\alpha=1,\ 2,\ 3)$$

$$\zeta_\alpha\,(u)=\zeta\,(u+\omega_\alpha)-\eta_\alpha \qquad (\sigma=1,\ 2,\ 3)$$

VI. Addition theorems of the Weierstrass functions

$$\wp(u) - \wp(v) = -\frac{\sigma(u+v)\,\sigma(u-v)}{\sigma^2(u)\,\sigma^2(v)}$$

$$\zeta(u+v) = \zeta(u) + \zeta(v) + \frac{1}{2}\frac{\wp'(u) - \wp'(v)}{\wp(u) - \wp(v)}$$

$$\wp(u+v) = \wp(u) - \frac{1}{2}\frac{\partial}{\partial u}\left\{\frac{\wp'(u) - \wp'(v)}{\wp(u) - \wp(v)}\right\}$$

$$\wp(u+v) + \wp(u) + \wp(v) = \frac{1}{4}\left\{\frac{\wp'(u) - \wp'(v)}{\wp(u) - \wp(v)}\right\}^2$$

$$\wp(u+v) - \wp(u-v) = -\frac{\wp'(u)\,\wp'(v)}{[\wp(u) - \wp(v)]^2}$$

$$\begin{vmatrix} 1 & \wp(u) & \wp'(u) \\ 1 & \wp(v) & \wp'(v) \\ 1 & \wp(w) & \wp'(w) \end{vmatrix} = 0 \qquad (u+v+w=0)$$

$$\wp(u+\omega_\alpha) - e_\alpha = \frac{(e_\alpha - e_\beta)(e_\alpha - e_\gamma)}{\wp(u) - e_\alpha} \qquad (\alpha,\ \beta,\ \gamma = 1,\ 2,\ 3)$$

$$\sqrt{\wp(u) - e_\alpha} = \frac{\sigma_\alpha(u)}{\sigma(u)} \qquad (\alpha = 1,\ 2,\ 3)$$

$$\wp'(u) = -2\frac{\sigma_1(u)\,\sigma_2(u)\,\sigma_3(u)}{[\sigma(u)]^3}$$

VII. Degeneration of the Weierstrass functions

$\omega' = \infty$, ω finite

$$\sigma(u) = \frac{2\omega}{\pi}\, e^{\frac{1}{3!}\left(\frac{\pi u}{2\omega}\right)^2} \sin\frac{\pi u}{2\omega}$$

$$\zeta(u) = \frac{1}{3}\left(\frac{\pi}{2\omega}\right)^2 u + \frac{\pi}{2\omega}\,\operatorname{ctg}\frac{\pi u}{2\omega}$$

$$\wp(u) = -\frac{1}{3}\left(\frac{\pi}{2\omega}\right)^2 + \left(\frac{\pi}{2\omega}\right)^2 \frac{1}{\sin^2\frac{\pi u}{2\omega}}$$

$$g_2^3 - 27 g_3^2 = 0$$

$$e_1 = \frac{3g_3}{g_2}\,, \quad e_2 = e_3 = -\frac{3g_3}{2g_2}$$

$$\left(\frac{\pi}{2\omega}\right)^2 = \frac{9g_3}{2g_2}\,, \quad 2\eta\omega = \frac{\pi^2}{6}$$

$\omega = \infty$, $\omega' = \infty$

$$\sigma(u) = u$$

$$\zeta(u) = \frac{1}{u}$$

$$\wp(u) = \frac{1}{u^2}$$

$$g_2 = g_3 = 0$$

$$e_1 = e_2 = e_3 = 0$$

VIII. Theta functions. Reduction formulas

$$h=e^{\pi i\tau}, \quad h^{\frac{1}{4}}=e^{\frac{\pi i\tau}{4}}, \quad \Im\tau>0, \quad z=e^{\pi iv}$$

$$\vartheta_1(v)=2h^{\frac{1}{4}}\sin\pi v-2h^{\frac{9}{4}}\sin 3\pi v+2h^{\frac{25}{4}}\sin 5\pi v-\ldots$$

$$\vartheta_2(v)=2h^{\frac{1}{4}}\cos\pi v+2h^{\frac{9}{4}}\cos 3\pi v+2h^{\frac{25}{4}}\cos 5\pi v+\ldots$$

$$\vartheta_3(v)=1+2h\cos 2\pi v+2h^4\cos 4\pi v+2h^9\cos 6\pi v+\ldots$$

$$\vartheta_0(v)=1-2h\cos 2\pi v+2h^4\cos 4\pi v-2h^9\cos 6\pi v+\ldots$$

$\vartheta_1(v\pm 1)=-\vartheta_1(v)$	$\vartheta_1\left(v\pm\dfrac{1}{2}\right)=\pm\,\vartheta_2(v)$
$\vartheta_2(v\pm 1)=-\vartheta_2(v)$	$\vartheta_2\left(v\pm\dfrac{1}{2}\right)=\mp\,\vartheta_1(v)$
$\vartheta_3(v\pm 1)=\vartheta_3(v)$	$\vartheta_3\left(v\pm\dfrac{1}{2}\right)=\vartheta_0(v)$
$\vartheta_0(v\pm 1)=\vartheta_0(v)$	$\vartheta_0\left(v\pm\dfrac{1}{2}\right)=\vartheta_3(v)$

$\vartheta_1(v\pm\tau)=-h^{-1}z^{\mp 2}\vartheta_1(v)$	$\vartheta_1\left(v\pm\dfrac{\tau}{2}\right)=\pm\,ih^{-\frac{1}{4}}z^{\mp 1}\vartheta_0(v)$
$\vartheta_2(v\pm\tau)=h^{-1}z^{\mp 2}\vartheta_2(v)$	$\vartheta_2\left(v\pm\dfrac{\tau}{2}\right)=h^{-\frac{1}{4}}z^{\mp 1}\vartheta_3(v)$
$\vartheta_3(v\pm\tau)=h^{-1}z^{\mp 2}\vartheta_3(v)$	$\vartheta_3\left(v\pm\dfrac{\tau}{2}\right)=h^{-\frac{1}{4}}z^{\mp 1}\vartheta_2(v)$
$\vartheta_0(v\pm\tau)=-h^{-1}z^{\mp 2}\vartheta_0(v)$	$\vartheta_0\left(v\pm\dfrac{\tau}{2}\right)=\pm\,ih^{-\frac{1}{4}}z^{\mp 1}\vartheta_1(v)$

All theta functions satisfy the differential equation

$$\frac{\partial^2\vartheta}{\partial v^2}=4\pi i\,\frac{\partial\vartheta}{\partial\tau} \qquad \{\vartheta=\vartheta(v\mid\tau)\}$$

IX. Expansion of Theta Functions in Infinite Products

$$h = e^{\pi i \tau}, \quad h^{\frac{1}{4}} = e^{\frac{1}{4}\pi i \tau}, \quad \Im \tau > 0$$

$$H_0 = \prod_{h=1}^{\infty} (1 - h^{2k}) \qquad H_1 = \prod_{k=1}^{\infty} (1 + h^{2k})$$

$$H_2 = \prod_{k=1}^{\infty} (1 + h^{2k-1}) \qquad H_3 = \prod_{k=1}^{\infty} (1 - h^{2k-1})$$

$$H_1 H_2 H_3 = 1$$

$$\vartheta_1(v) = 2H_0 h^{\frac{1}{4}} \sin \pi v \prod_{k=1}^{\infty} (1 - 2h^{2k} \cos 2\pi v + h^{4k})$$

$$\vartheta_2(v) = 2H_0 h^{\frac{1}{4}} \cos \pi v \prod_{k=1}^{\infty} (1 + 2h^{2k} \cos 2\pi v + h^{4k})$$

$$\vartheta_3(v) = H_0 \prod_{k=1}^{\infty} (1 + 2h^{2k-1} \cos 2\pi v + h^{4k-2})$$

$$\vartheta_0(v) = H_0 \prod_{k=1}^{\infty} (1 - 2h^{2k-1} \cos 2\pi v + h^{4k-2})$$

Zeros of theta functions

$\vartheta_1(v)$	$\vartheta_2(v)$	$\vartheta_3(v)$	$\vartheta_0(v)$
$m + n\tau$	$m - \dfrac{1}{2} + n\tau$	$m - \dfrac{1}{2} + \left(n - \dfrac{1}{2}\right)\tau$	$m + \left(n - \dfrac{1}{2}\right)\tau$

$$(m, \; n = 0, \; \pm 1, \; \pm 2, \; \ldots)$$

Zero values of theta functions

$\vartheta_1' = \vartheta_1'(0) = 2\pi h^{\frac{1}{4}} H_0^3$ $\vartheta_3 = \vartheta_3(0) = H_0 H_2^2$	$\vartheta_2 = \vartheta_2(0) = 2h^{\frac{1}{4}} H_0 H_1^2$ $\vartheta_0 = \vartheta_0(0) = H_0 H_3^2$
$\vartheta_1' = \pi \vartheta_2 \vartheta_3 \vartheta_0$	$\vartheta_3^4 = \vartheta_0^4 + \vartheta_2^4$

X. Various Expansions in Simple Series

$\dfrac{\omega'}{\omega}=\tau$	$\Im\tau>0$	$h=e^{\pi i\tau}$	$v=\dfrac{u}{2\omega}$	$e^{\pi i v}=z$

$$\sigma(u)=2\omega e^{2\eta\omega v^2}\,\frac{\vartheta_1(v)}{\vartheta_1'}\qquad\qquad \sigma_1(u)=e^{2\eta\omega v^2}\,\frac{\vartheta_2(v)}{\vartheta_2}$$

$$\sigma_2(u)=e^{2\eta\omega v^2}\,\frac{\vartheta_3(v)}{\vartheta_3}\qquad\qquad \sigma_3(u)=e^{2\eta\omega v^2}\,\frac{\vartheta_0(v)}{\vartheta_0}$$

$$\zeta(u)=\frac{\eta}{\omega}u+\frac{\pi i}{2\omega}\left\{\frac{z+z^{-1}}{z-z^{-1}}+\sum_{k=1}^{\infty}\frac{2h^{2k}z^{-2}}{1-h^{2k}z^{-2}}-\sum_{k=1}^{\infty}\frac{2h^{2k}z^2}{1-h^{2k}z^2}\right\}$$

$$\wp(u)=-\frac{\eta}{\omega}-\left(\frac{\pi}{\omega}\right)^2\left\{\frac{1}{(z-z^{-1})^2}+\sum_{k=1}^{\infty}\frac{h^{2k}z^{-2}}{(1-h^{2k}z^{-2})^2}+\sum_{k=1}^{\infty}\frac{h^{2k}z^2}{(1-h^{2k}z^2)^2}\right\}$$

$$2\eta\omega=-\frac{1}{6}\frac{\vartheta_1'''}{\vartheta_1'}=\frac{\pi^2}{6}\frac{1-3^3h^{1\times2}+5^3h^{2\times3}-\ldots}{1-3h^{1\times2}+5h^{2\times3}-\ldots}=$$

$$=\frac{\pi^2}{6}\left\{1-24\sum_{k=1}^{\infty}\frac{h^{2k}}{(1-h^{2k})^2}\right\}$$

$$e_1=-\frac{\eta}{\omega}+\left(\frac{\pi}{\omega}\right)^2\left\{\frac{1}{4}+2\sum_{k=1}^{\infty}\frac{h^{2k}}{(1+h^{2k})^2}\right\}$$

$$e_2=-\frac{\eta}{\omega}+2\left(\frac{\pi}{\omega}\right)^2\sum_{k=1}^{\infty}\frac{h^{2k-1}}{(1+h^{2k-1})^2}$$

$$e_3=-\frac{\eta}{\omega}-2\left(\frac{\pi}{\omega}\right)^2\sum_{k=1}^{\infty}\frac{h^{2k-1}}{(1-h^{2k-1})^2}$$

$$\sqrt{e_2-e_1}=i\sqrt{e_1-e_2}=i\frac{\pi}{2\omega}\vartheta_0^2$$

$$\sqrt{e_3-e_2}=i\sqrt{e_2-e_3}=-i\frac{\pi}{2\omega}\vartheta_2^2$$

$$\sqrt{e_1-e_3}=i\sqrt{e_3-e_1}=\frac{\pi}{2\omega}\vartheta_3^2$$

XI. OTHER NOTATIONS FOR THETA FUNCTIONS

$h = e^{\pi i \tau}$	$\Im \tau > 0$

$$K = \frac{\pi}{2}\,\{1 + 2h + 2h^4 + \ldots\}^2 = \frac{\pi}{2}\,\vartheta_3^2 = \frac{\pi}{2}\,\vartheta_3^2\,(0\mid\tau)$$

$$iK' = \tau K$$

$\lambda = e^{-\frac{\pi i}{4K}(2u+iK')}$	$\mu = e^{-\frac{\pi i}{K}(u+iK')}$

$$0_\alpha\,(w) = \vartheta_\alpha\left(\frac{w}{2K}\right) \qquad (\alpha = 0,\ 1,\ 2,\ 3)$$

$H\,(w) = 0_1\,(w)$ $H_1\,(w) = 0_2\,(w)$	$\Theta\,(w) = 0_0\,(w)$ $\Theta_1\,(w) = 0_3\,(w)$

Zeros

$H\,(w)$	$H_1\,(w)$	$\Theta\,(w)$	$\Theta_1\,(w)$
$2mK + 2niK'$	$(2m+1)\,K + 2niK'$	$2mK + (2n+1)\,iK'$	$(2m+1)\,K + (2n+1)\,iK'$

$$(m,\ n = 0,\ \pm 1,\ \pm 2,\ \ldots)$$

$H\,(u+K) = H_1\,(u)$	$H\,(u+iK') = i\lambda\,\Theta\,(u)$
$\Theta\,(u+K) = \Theta_1\,(u)$	$\Theta\,(u+iK') = i\lambda\,H\,(u)$
$H_1\,(u+K) = -H\,(u)$	$H_1\,(u+iK') = \lambda\,\Theta_1\,(u)$
$\Theta_1\,(u+K) = \Theta\,(u)$	$\Theta_1\,(u+iK') = \lambda\,H_1\,(u)$

$H\,(u+K+iK') = \lambda\,\Theta_1\,(u)$	$H\,(u+2iK') = -\mu\,H\,(u)$
$\Theta\,(u+K+iK') = \lambda\,H_1\,(u)$	$\Theta\,(u+2iK') = -\mu\,\Theta\,(u)$
$H_1\,(u+K+iK') = -i\lambda\,\Theta\,(u)$	$H_1\,(u+2iK') = \mu\,H_1\,(u)$
$\Theta_1\,(u+K+iK') = i\lambda\,H\,(u)$	$\Theta_1\,(u+2iK') = \mu\,\Theta_1\,(u)$

XII. Jacobi functions

$$\sqrt{\frac{2K}{\pi}} = 1 + 2h + 2h^4 + 2h^9 + \cdots$$

$$\sqrt{k} = \frac{H_1(0)}{\Theta_1(0)} = \frac{2h^{\frac{1}{4}} + 2h^{\frac{9}{4}} + \cdots}{1 + 2h + 2h^4 + \cdots} \qquad \left(h^{\frac{1}{4}} = e^{\frac{1}{4}\pi i \tau}\right)$$

$$\sqrt{k'} = \frac{\Theta(0)}{\Theta_1(0)} = \frac{1 - 2h + 2h^4 - \cdots}{1 + 2h + 2h^4 + \cdots}$$

$$\mathrm{sn}\,(u;\,k) = \frac{1}{\sqrt{k}}\,\frac{H(u)}{\Theta(u)}$$

$$\mathrm{cn}\,(u;\,k) = \sqrt{\frac{k'}{k}}\,\frac{H_1(u)}{\Theta(u)}$$

$$\mathrm{dn}\,(u;\,k) = \sqrt{k'}\,\frac{\Theta_1(u)}{\Theta(u)}$$

$$\begin{array}{ll}
\mathrm{sn}\,(u+2K) = -\,\mathrm{sn}\,u & \mathrm{sn}\,(u+2iK') = \mathrm{sn}\,u \\
\mathrm{cn}\,(u+2K) = -\,\mathrm{cn}\,u & \mathrm{cn}\,(u+2iK') = -\,\mathrm{cn}\,u \\
\mathrm{dn}\,(u+2K) = \mathrm{dn}\,u & \mathrm{dn}\,(u+2iK') = -\,\mathrm{dn}\,u
\end{array}$$

Periods

$\mathrm{sn}\,u$	$\mathrm{cn}\,u$	$\mathrm{dn}\,u$
$4K,\ 2iK'$	$4K,\ 2K+2iK'$	$2K,\ 4iK'$

$$\begin{array}{ll}
\mathrm{sn}\,(u+K) = \dfrac{\mathrm{cn}\,u}{\mathrm{dn}\,u} & \mathrm{sn}\,(u+iK') = \dfrac{1}{k\,\mathrm{sn}\,u} \\[2ex]
\mathrm{cn}\,(u+K) = -\,k'\,\dfrac{\mathrm{sn}\,u}{\mathrm{dn}\,u} & \mathrm{cn}\,(u+iK') = -\,i\,\dfrac{\mathrm{dn}\,u}{k\,\mathrm{sn}\,u} \\[2ex]
\mathrm{dn}\,(u+K) = \dfrac{k'}{\mathrm{dn}\,u} & \mathrm{dn}\,(u+iK') = -\,i\,\dfrac{\mathrm{cn}\,u}{\mathrm{sn}\,u}
\end{array}$$

$$\mathrm{sn}\,(u+K+iK') = \frac{1}{k}\,\frac{\mathrm{dn}\,u}{\mathrm{cn}\,u}$$

$$\mathrm{cn}\,(u+K+iK') = -\,\frac{ik'}{k\,\mathrm{cn}\,u}$$

$$\mathrm{dn}\,(u+K+iK') = ik'\,\frac{\mathrm{sn}\,u}{\mathrm{cn}\,u}$$

XIII. Some values of Jacobi functions

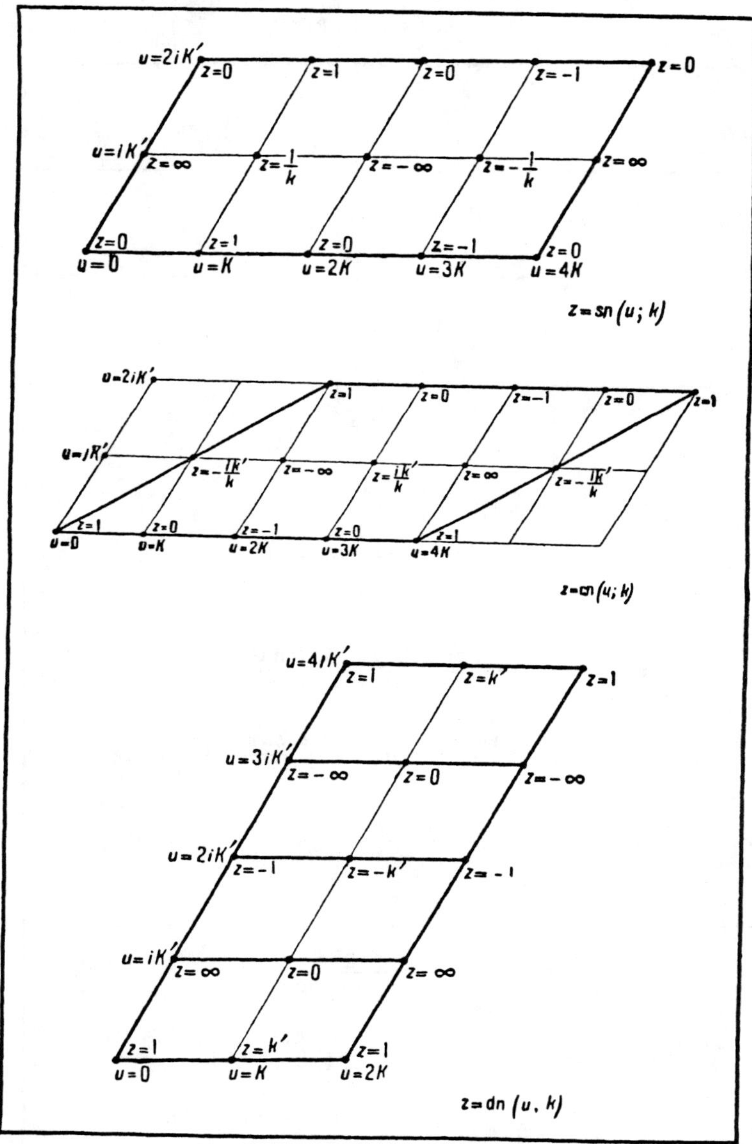

XIV. Differentiation of Jacobi functions. Addition theorems

$$z = \operatorname{sn} u \qquad z'^2 = (1 - z^2)(1 - k^2 z^2)$$

$$z = \operatorname{cn} u \qquad z'^2 = (1 - z^2)(1 - k^2 + k^2 z^2)$$

$$z = \operatorname{dn} u \qquad z'^2 = (1 - z^2)(z^2 - 1 + k^2)$$

$$\frac{d}{du} \operatorname{sn} u = \operatorname{cn} u \; \operatorname{dn} u$$

$$\frac{d}{du} \operatorname{cn} u = -\operatorname{sn} u \; \operatorname{dn} u$$

$$\frac{d}{du} \operatorname{dn} u = -k^2 \operatorname{sn} u \; \operatorname{cn} u$$

$$\operatorname{sn}(u + v) = \frac{\operatorname{sn} u \, \operatorname{cn} v \, \operatorname{dn} v + \operatorname{sn} v \, \operatorname{cn} u \, \operatorname{dn} u}{1 - k^2 \operatorname{sn}^2 u \, \operatorname{sn}^2 v}$$

$$\operatorname{cn}(u + v) = \frac{\operatorname{cn} u \, \operatorname{cn} v - \operatorname{sn} u \, \operatorname{dn} u \, \operatorname{sn} v \, \operatorname{dn} v}{1 - k^2 \operatorname{sn}^2 u \, \operatorname{sn}^2 v}$$

$$\operatorname{dn}(u + v) = \frac{\operatorname{dn} u \, \operatorname{dn} v - k^2 \operatorname{sn} u \, \operatorname{cn} u \, \operatorname{sn} v \, \operatorname{cn} v}{1 - k^2 \operatorname{sn}^2 u \, \operatorname{sn}^2 v}$$

$$\operatorname{sn}(u + v)\operatorname{sn}(u - v) = \frac{\operatorname{sn}^2 u - \operatorname{sn}^2 v}{1 - k^2 \operatorname{sn}^2 u \, \operatorname{sn}^2 v}$$

$$\operatorname{cn}(u + v)\operatorname{cn}(u - v) = \frac{\operatorname{cn}^2 v - \operatorname{dn}^2 v \, \operatorname{sn}^2 u}{1 - k^2 \operatorname{sn}^2 u \, \operatorname{sn}^2 v}$$

$$\operatorname{dn}(u + v)\operatorname{dn}(u - v) = \frac{\operatorname{dn}^2 v - k^2 \operatorname{cn}^2 v \, \operatorname{sn}^2 u}{1 - k^2 \operatorname{sn}^2 u \, \operatorname{sn}^2 v}$$

XV. Certain values of the Jacobi functions (continuation)

$$\operatorname{sn} \frac{K}{2} = \frac{1}{\sqrt{1+k'}}$$

$$\operatorname{cn} \frac{K}{2} = \frac{\sqrt{k'}}{\sqrt{1+k'}}$$

$$\operatorname{dn} \frac{K}{2} = \sqrt{k'}$$

$$\operatorname{sn} \frac{iK'}{2} = \frac{i}{\sqrt{k}}$$

$$\operatorname{cn} \frac{iK'}{2} = \frac{\sqrt{1+k}}{\sqrt{k}}$$

$$\operatorname{dn} \frac{iK'}{2} = \sqrt{1+k}$$

$$\operatorname{sn} \frac{K+iK'}{2} = \frac{1}{\sqrt{2}\,\sqrt{k}}\left(\sqrt{1+k}+i\sqrt{1-k}\right)$$

$$\operatorname{cn} \frac{K+iK'}{2} = \frac{\sqrt{k'}\,(1-i)}{\sqrt{2}\,\sqrt{k}}$$

$$\operatorname{dn} \frac{K+iK'}{2} = \frac{\sqrt{k'}}{\sqrt{2}}\left(\sqrt{1+k'}-i\sqrt{1-k'}\right)$$

In the normal case $(0 < k < 1)$ all the roots are arithmetic

XVI. Elliptic Integrals of the First and Second Kinds

If the points e_1, e_2, and e_3 lie on a single line, then e_2 denotes the middle one of these points.

$$k^2 = \frac{e_2 - e_3}{e_1 - e_3} \qquad\qquad k'^2 = \frac{e_1 - e_2}{e_1 - e_3}$$

$$K = \int_0^1 \frac{dt}{\sqrt{(1-t^2)(1-k^2 t^2)}} \qquad K' = \int_0^1 \frac{dt}{\sqrt{(1-t^2)(1-k'^2 t^2)}}$$

$$E = \int_0^1 \sqrt{\frac{1-k^2 t^2}{1-t^2}}\, dt \qquad E' = \int_0^1 \sqrt{\frac{1-k'^2 t^2}{1-t^2}}\, dt$$

The integral is taken along a rectilinear path with a positive circuit over a small semicircle about the point $t = 1/|k|$ if $1 < k^2 < \infty$, and about the point $t = 1/|k'|$ if $1 < k'^2 < \infty$. In the normal case $(0 < k < 1)$ all the integrals are positive.)

$$\eta_1 = \sqrt{e_1 - e_3}\left\{ E - \frac{e_1}{e_1 - e_3} K \right\} \qquad \eta_3 = -i\sqrt{e_1 - e_3}\left\{ E' + \frac{e_3}{e_1 - e_3} K' \right\}$$

$$EK' + E'K - KK' = \frac{1}{2}\pi$$

$$\omega_1 = \frac{K}{\sqrt{e_1 - e_3}} \qquad\qquad \omega_3 = \frac{iK'}{\sqrt{e_1 - e_3}}$$

$$\frac{\sigma(u)}{\sigma_3(u)} = \frac{1}{\sqrt{e_1 - e_3}}\, \mathrm{sn}\,(\sqrt{e_1 - e_3}\, u;\, k)$$

$$\frac{\sigma_1(u)}{\sigma_3(u)} = \mathrm{cn}\,(\sqrt{e_1 - e_3}\, u;\, k) \qquad \frac{\sigma_2(u)}{\sigma_3(u)} = \mathrm{dn}\,(\sqrt{e_1 - e_3}\, u;\, k)$$

$$E(u) = \int_0^u \mathrm{dn}^2 v\, dv \qquad Z(u) = E(u) - \frac{E}{K}\, u = \frac{\Theta'(u)}{\Theta(u)}$$

$$\int \frac{du}{\mathrm{sn}^2 u} = u\, Z'(0) - \frac{H'(u)}{H(u)} \qquad \int \frac{du}{\mathrm{cn}^2 u} = \frac{u}{k'^2}\, Z'(K) - \frac{1}{k'^2}\, \frac{H_1'(u)}{H_1(u)}$$

$$\int \frac{du}{\mathrm{dn}^2 u} = \frac{u}{k'^2}\, \frac{E}{K} + \frac{1}{k'^2}\, \frac{\Theta_1'(u)}{\Theta_1(u)}$$

XVII. Transformation of theta functions
(of the first degree)

$$h = e^{\pi i \tau}, \ \Im \tau > 0, \ \tau' = -\frac{1}{\tau}, \ \Re \sqrt{-i\tau'} > 0, \ i^{-\frac{1}{2}} = e^{-\frac{\pi i}{4}}$$

$$\vartheta_1(v \mid \tau) = i^{-\frac{1}{2}} \vartheta_1(v \mid \tau + 1)$$

$$\vartheta_2(v \mid \tau) = i^{-\frac{1}{2}} \vartheta_2(v \mid \tau + 1)$$
$$\vartheta_3(v \mid \tau) = \vartheta_0(v \mid \tau + 1)$$
$$\vartheta_0(v \mid \tau) = \vartheta_3(v \mid \tau + 1)$$

$$\vartheta_1(v \mid \tau) = -i \sqrt{-i\tau'} \, e^{\tau' \pi i v^2} \vartheta_1(\tau'v \mid \tau')$$
$$\vartheta_2(v \mid \tau) = \sqrt{-i\tau'} \, e^{\tau' \pi i v^2} \vartheta_0(\tau'v \mid \tau')$$
$$\vartheta_3(v \mid \tau) = \sqrt{-i\tau'} \, e^{\tau' \pi i v^2} \vartheta_3(\tau'v \mid \tau')$$
$$\vartheta_0(v \mid \tau) = \sqrt{-i\tau'} \, e^{\tau' \pi i v^2} \vartheta_2(\tau'v \mid \tau')$$

XVIII. The first principal
FIRST-DEGREE TRANSFORMATION

$$\lambda = \frac{ik}{k'} \qquad M = \frac{1}{k'}$$

$$L = \frac{K}{M} \qquad iL' = \frac{iK' + K}{M}$$

$$\operatorname{sn}\left(\frac{u}{M} ; \lambda\right) = \frac{1}{M} \frac{\operatorname{sn}(u; k)}{\operatorname{dn}(u; k)}$$

$$\operatorname{cn}\left(\frac{u}{M} ; \lambda\right) = \frac{\operatorname{cn}(u; k)}{\operatorname{dn}(u; k)}$$

$$\operatorname{dn}\left(\frac{u}{M} ; \lambda\right) = \frac{1}{\operatorname{dn}(u; k)}$$

XIX. The second principal
FIRST-DEGREE TRANSFORMATION

$$\lambda = k' \qquad M = \frac{1}{i}$$

$$L = \frac{iK'}{M} \qquad iL' = -\frac{K}{M}$$

$$\operatorname{sn}\left(\frac{u}{M} ; \lambda\right) = \frac{1}{M} \frac{\operatorname{sn}(u; k)}{\operatorname{cn}(u; k)}$$

$$\operatorname{cn}\left(\frac{u}{M} ; \lambda\right) = \frac{1}{\operatorname{cn}(u; k)}$$

$$\operatorname{dn}\left(\frac{u}{M} ; \lambda\right) = \frac{\operatorname{dn}(u; k)}{\operatorname{cn}(u; k)}$$

XX. Landen's Transformation

$$\lambda = \frac{1-k'}{1+k'} \qquad M = \frac{1}{1+k'}$$

$$L = \frac{K}{2M} \qquad L' = \frac{K'}{M}$$

$$\operatorname{sn}\left(\frac{u}{M}\,;\,\lambda\right) = \frac{1}{M}\,\frac{\operatorname{sn}(u;\,k)\,\operatorname{cn}(u;\,k)}{\operatorname{dn}(u;\,k)}$$

$$\operatorname{cn}\left(\frac{u}{M}\,;\,\lambda\right) = \frac{1-(1+k')\operatorname{sn}^2(u;\,k)}{\operatorname{dn}(u;\,k)}$$

$$\operatorname{dn}\left(\frac{u}{M}\,;\,\lambda\right) = \frac{1-(1-k')\operatorname{sn}^2(u;\,k)}{\operatorname{dn}(u;\,k)}$$

XXI. Gauss's Transformation

$$\lambda = \frac{2\sqrt{k}}{1+k} \qquad M = \frac{1}{1+k}$$

$$L = \frac{K}{M} \qquad L' = \frac{K'}{2M}$$

$$\operatorname{sn}\left(\frac{u}{M}\,;\,\lambda\right) = \frac{1}{M}\,\frac{\operatorname{sn}(u;\,k)}{1+k\operatorname{sn}^2(u;\,k)}$$

$$\operatorname{cn}\left(\frac{u}{M}\,;\,\lambda\right) = \frac{\operatorname{cn}(u;\,k)\,\operatorname{dn}(u;\,k)}{1+k\operatorname{sn}^2(u;\,k)}$$

$$\operatorname{dn}\left(\frac{u}{M}\,;\,\lambda\right) = \frac{1-k\operatorname{sn}^2(u;\,k)}{1+k\operatorname{sn}^2(u;\,k)}$$

$$\vartheta_1\left(v\left|\frac{\tau}{2}\right.\right) = \frac{2\vartheta_1(v\,|\,\tau)\,\vartheta_0(v\,|\,\tau)}{\vartheta_2\left(0\left|\frac{\tau}{2}\right.\right)}$$

$$\vartheta_2\left(v\left|\frac{\tau}{2}\right.\right) = \frac{2\vartheta_2(v\,|\,\tau)\,\vartheta_3(v\,|\,\tau)}{\vartheta_2\left(0\left|\frac{\tau}{2}\right.\right)}$$

$$\vartheta_3\left(v\left|\frac{\tau}{2}\right.\right) = \frac{\vartheta_0^2(v\,|\,\tau)-\vartheta_1^2(v\,|\,\tau)}{\vartheta_0\left(0\left|\frac{\tau}{2}\right.\right)}$$

$$\vartheta_0\left(v\left|\frac{\tau}{2}\right.\right) = \frac{\vartheta_0^2(v\,|\,\tau)+\vartheta_1^2(v\,|\,\tau)}{\vartheta_3\left(0\left|\frac{\tau}{2}\right.\right)}$$

XXII. The first principal nth-degree transformation

$$L = \frac{K}{nM}, \quad L' = \frac{K'}{M}$$

$$c_r = \mathrm{sn}^2\left(\frac{rK}{n}\ ;\ k\right)$$

$$\lambda = k^n \prod_{r=1}^{[\frac{n}{2}]} c_{2r-1}^2 \qquad M = \prod_{r=1}^{[\frac{n}{2}]} \frac{c_{2r-1}}{c_{2r}}$$

A. n odd

$$\mathrm{sn}\left(\frac{u}{M}\ ;\ \lambda\right) = \frac{1}{M}\,\mathrm{sn}(u;\ k) \prod_{r=1}^{\frac{n-1}{2}} \frac{1 - \dfrac{\mathrm{sn}^2(u;\ k)}{c_{2r}}}{1 - k^2 c_{2r}\,\mathrm{sn}^2(u;\ k)}$$

$$\mathrm{cn}\left(\frac{u}{M}\ ;\ \lambda\right) = \mathrm{cn}(u;\ k) \prod_{r=1}^{\frac{n-1}{2}} \frac{1 - \dfrac{\mathrm{sn}^2(u;\ k)}{c_{2r-1}}}{1 - k^2 c_{2r}\,\mathrm{sn}^2(u;\ k)}$$

$$\mathrm{dn}\left(\frac{u}{M}\ ;\ \lambda\right) = \mathrm{dn}(u;\ k) \prod_{r=1}^{\frac{n-1}{2}} \frac{1 - k^2 c_{2r-1}\,\mathrm{sn}^2(u;\ k)}{1 - k^2 c_{2r}\,\mathrm{sn}^2(u;\ k)}$$

B. n even

$$\mathrm{sn}\left(\frac{u}{M} + L;\ \lambda\right) = \prod_{r=1}^{\frac{n}{2}} \frac{1 - \dfrac{\mathrm{sn}^2(u;\ k)}{c_{2r-1}}}{1 - k^2 c_{2r-1}\,\mathrm{sn}^2(u;\ k)}$$

$$\mathrm{cn}\left(\frac{u}{M} + L;\ \lambda\right) = -\frac{\lambda'}{M}\,\frac{\mathrm{sn}(u;\ k)}{\mathrm{cn}(u;\ k)} \prod_{r=1}^{\frac{n}{2}} \frac{1 - \dfrac{\mathrm{sn}^2(u;\ k)}{c_{2r}}}{1 - k^2 c_{2r-1}\,\mathrm{sn}^2(u;\ k)}$$

$$\mathrm{dn}\left(\frac{u}{M} + L;\ \lambda\right) = \frac{\lambda'}{\mathrm{dn}(u;\ k)} \prod_{r=1}^{\frac{n}{2}} \frac{1 - k^2 c_{2r}\,\mathrm{sn}^2(u;\ k)}{1 - k^2 c_{2r-1}\,\mathrm{sn}^2(u;\ k)}$$

XXIII. The Second Principal nTH-Degree Transformation

$$L=\frac{K}{M}, \quad L'=\frac{K'}{nM}$$

$$\lambda=\prod_{r=1}^{n}\frac{\Theta^2\left(\frac{2r}{n}K';k'\right)}{\Theta^2\left(\frac{2r-1}{n}K';k'\right)} \qquad M=\prod_{r=1}^{\left[\frac{n}{2}\right]}\frac{\mathrm{sn}^2\left(\frac{2r-1}{n}K',k'\right)}{\mathrm{sn}^2\left(\frac{2r}{n}K';k'\right)}$$

$$c_r=\frac{\mathrm{sn}^2\left(\frac{rK'}{n};k'\right)}{\mathrm{cn}^2\left(\frac{rK'}{n};k'\right)} \qquad \delta_r=\mathrm{dn}^2\left(\frac{rK'}{n};k'\right)$$

$$\mathrm{sn}\left(\frac{u}{M};\lambda\right)=\frac{1}{M}\,\mathrm{sn}(u;k)\prod_{r=1}^{\left[\frac{n}{2}\right]}\frac{1+\dfrac{\mathrm{sn}^2(u;k)}{c_{2r}}}{1+\dfrac{\mathrm{sn}^2(u;k)}{c_{2r-1}}}$$

A. n odd

$$\mathrm{cn}\left(\frac{u}{M};\lambda\right)=\mathrm{cn}(u;k)\prod_{r=1}^{\frac{n-1}{2}}\frac{1-\delta_{2r}\,\mathrm{sn}^2(u;k)}{1+\dfrac{\mathrm{sn}^2(u;k)}{c_{2r-1}}}$$

$$\mathrm{dn}\left(\frac{u}{M};\lambda\right)=\mathrm{dn}(u;k)\prod_{r=1}^{\frac{n-1}{2}}\frac{1-\delta_{2r-1}\,\mathrm{sn}^2(u;k)}{1+\dfrac{\mathrm{sn}^2(u;k)}{c_{2r-1}}}$$

B. n even

$$\mathrm{cn}\left(\frac{u}{M};\lambda\right)=\mathrm{cn}(u;k)\,\mathrm{dn}(u;k)\,\frac{\displaystyle\prod_{r=1}^{\frac{n}{2}-1}1-\delta_{2r}\,\mathrm{sn}^2(u;k)}{\displaystyle\prod_{r=1}^{\frac{n}{2}}1+\dfrac{\mathrm{sn}^2(u;k)}{c_{2r-1}}}$$

$$\mathrm{dn}\left(\frac{u}{M};\lambda\right)=\prod_{r=1}^{\frac{n}{2}}\frac{1-\delta_{2r-1}\,\mathrm{sn}^2(u;k)}{1+\dfrac{\mathrm{sn}^2(u;k)}{c_{2r-1}}}$$

XXIV. Some integrals

$$\int \operatorname{sn} u \, du = -\frac{1}{k} \ln \left(\operatorname{dn} u + k \operatorname{cn} u \right)$$

$$\int \operatorname{cn} u \, du = \frac{i}{k} \ln \left(\operatorname{dn} u - ik \operatorname{sn} u \right)$$

$$\int \operatorname{dn} u \, du = i \ln \left(\operatorname{cn} u - i \operatorname{sn} u \right)$$

$$\int \frac{du}{\operatorname{sn} u} = \ln \frac{\operatorname{dn} u - \operatorname{cn} u}{\operatorname{sn} u}$$

$$\int \frac{du}{\operatorname{cn} u} = \frac{1}{k'} \ln \frac{\operatorname{dn} u + k' \operatorname{sn} u}{\operatorname{cn} u}$$

$$\int \frac{du}{\operatorname{dn} u} = \frac{1}{ik'} \ln \frac{\operatorname{cn} u + ik' \operatorname{sn} u}{\operatorname{dn} u}$$

$$\int \frac{\operatorname{sn} u}{\operatorname{cn} u} \, du = \frac{1}{k'} \ln \frac{\operatorname{dn} u + k'}{\operatorname{cn} u}$$

$$\int \frac{\operatorname{cn} u}{\operatorname{dn} u} \, du = -\frac{1}{k} \ln \frac{1 - k \operatorname{sn} u}{\operatorname{dn} u}$$

$$\int \frac{\operatorname{dn} u}{\operatorname{sn} u} \, du = \ln \frac{1 - \operatorname{cn} u}{\operatorname{sn} u}$$

$$\int \frac{\operatorname{sn} u}{\operatorname{dn} u} \, du = \frac{i}{kk'} \ln \frac{ik' - k \operatorname{cn} u}{\operatorname{dn} u}$$

$$\int \frac{\operatorname{cn} u}{\operatorname{sn} u} \, du = \ln \frac{1 - \operatorname{dn} u}{\operatorname{sn} u}$$

$$\int \frac{\operatorname{dn} u}{\operatorname{cn} u} \, du = \ln \frac{1 + \operatorname{sn} u}{\operatorname{cn} u}$$

$$\int \frac{\operatorname{sn} u}{\operatorname{cn}^2 u} \, du = \frac{1}{k'^2} \frac{\operatorname{dn} u}{\operatorname{cn} u}$$

$$\int \frac{\operatorname{sn} u}{\operatorname{dn}^2 u} \, du = -\frac{1}{k'^2} \frac{\operatorname{cn} u}{\operatorname{dn} u}$$

$$\int \frac{\operatorname{cn} u}{\operatorname{sn}^2 u} \, du = -\frac{\operatorname{dn} u}{\operatorname{sn} u}$$

$$\int \frac{\operatorname{cn} u}{\operatorname{dn}^2 u} \, du = \frac{\operatorname{sn} u}{\operatorname{dn} u}$$

$$\int \frac{\operatorname{dn} u}{\operatorname{sn}^2 u} \, du = -\frac{\operatorname{cn} u}{\operatorname{sn} u}$$

$$\int \frac{\operatorname{dn} u}{\operatorname{cn}^2 u} \, du = \frac{\operatorname{sn} u}{\operatorname{cn} u}$$

XXV. Computation of Elliptic Integrals in the Real Case

$$\int R(x, y)\, dx$$

I	$y = \sqrt{(a^2 - x^2)(b^2 - x^2)} \qquad a^2 > b^2$ $k^2 = \dfrac{b^2}{a^2} \qquad x = \begin{cases} b\,\mathrm{sn}\,u & (x^2 < b^2) \\[2mm] \dfrac{a}{\mathrm{sn}\,u} & (x^2 > a^2) \end{cases}$
II	$y = \sqrt{(a^2 - x^2)(x^2 - b^2)} \qquad a^2 > b^2 \quad b^2 < x^2 < a^2$ $k^2 = \dfrac{a^2 - b^2}{a^2} \qquad x = a\,\mathrm{dn}\,u$
III	$y = \sqrt{(a^2 - x^2)(b^2 + x^2)} \qquad x^2 < a^2$ $k^2 = \dfrac{a^2}{a^2 + b^2} \qquad x = a\,\mathrm{cn}\,u$
IV	$y = \sqrt{(x^2 - a^2)(b^2 + x^2)} \qquad x^2 > a^2$ $k^2 = \dfrac{b^2}{a^2 + b^2} \qquad x = \dfrac{a}{\mathrm{cn}\,u}$
V	$y = \sqrt{(a^2 + x^2)(b^2 + x^2)} \qquad a^2 > b^2$ $k^2 = \dfrac{a^2 - b^2}{a^2} \qquad x = a\,\dfrac{\mathrm{cn}\,u}{\mathrm{sn}\,u}$

Tables of Values of Elliptic Integrals

I. Complete elliptic integrals

$$K = \int_0^1 \frac{dt}{\sqrt{(1-t^2)(1-k^2t^2)}}, \quad E = \int_0^1 \sqrt{\frac{1-k^2t^2}{1-t^2}}\, dt$$

$\alpha°$	$k^2 = \sin^2\alpha$	K	E	$\alpha°$	$k^2 = \sin^2\alpha$	K	E
0	0,00000	1,57080	1,57080	28	0,22040	1,67006	1,48029
1	0,00030	1,57092	1,57068	29	0,23504	1,67773	1,47397
2	0,00122	1,57127	1,57032	30	0,25000	1,68575	1,46746
3	0,00274	1,57187	1,56972	31	0,26526	1,69411	1,46077
4	0,00487	1,57271	1,56888	32	0,28081	1,70284	1,45391
5	0,00760	1,57379	1,56781	33	0,29663	1,71192	1,44687
6	0,01093	1,57511	1,56650	34	0,31270	1,72139	1,43966
7	0,01485	1,57668	1,56495	35	0,32899	1,73125	1,43229
8	0,01937	1,57849	1,56316	36	0,34549	1,74150	1,42476
9	0,02447	1,58054	1,56114	37	0,36218	1,75217	1,41707
10	0,03015	1,58284	1,55889	38	0,37904	1,76326	1,40924
11	0,03641	1,58539	1,55640	39	0,39604	1,77479	1,40126
12	0,04323	1,58820	1,55368	40	0,41318	1,78677	1,39314
13	0,05060	1,59125	1,55073	41	0,43041	1,79922	1,38489
14	0,05853	1,59457	1,54755	42	0,44774	1,81216	1,37650
15	0,06699	1,59814	1,54415	43	0,46512	1,82560	1,36800
16	0,07598	1,60198	1,54052	44	0,48255	1,83957	1,35938
17	0,08548	1,60608	1,53667	45	0,50000	1,85407	1,35064
18	0,09549	1,61045	1,53260	46	0,51745	1,86915	1,34181
19	0,10599	1,61510	1,52831	47	0,53488	1,88481	1,33287
20	0,11698	1,62003	1,52380	48	0,55226	1,90108	1,32384
21	0,12843	1,62523	1,51908	49	0,56959	1,91800	1,31473
22	0,14033	1,63073	1,51415	50	0,58682	1,93558	1,30554
23	0,15267	1,63652	1,50901	51	0,60396	1,95386	1,29628
24	0,16543	1,64260	1,50366	52	0,62096	1,97288	1,28695
25	0,17861	1,64900	1,49811	53	0,63782	1,99267	1,27757
26	0,19217	1,65570	1,49237	54	0,65451	2,01327	1,26815
27	0,20611	1,66272	1,48643	55	0,67101	2,03472	1,25868

TABLES OF VALUES OF ELLIPTIC INTEGRALS

CONTINUATION

$\alpha°$	$k^2 = \sin^2 \alpha$	K	E	$\alpha°$	$k^2 = \sin^2 \alpha$	K	E
56	0,68730	2,05706	1,24918	82,4	0,98251	3,41994	1,02558
57	0,70337	2,08036	1,23966	82,6	0,98341	3,44601	1,02447
58	0,71919	2,10466	1,23013	82,8	0,98429	3,47282	1,02338
59	0,73474	2,13002	1,22059	83,0	0,98515	3,50042	1,02231
60	0,75000	2,15652	1,21106	83,2	0,98598	3,52884	1,02126
61	0,76496	2,18421	1,20154	83,4	0,98680	3,55814	1,02023
62	0,77960	2,21319	1,19205	83,6	0,98757	3,58837	1,01921
63	0,79389	2,24355	1,18259	83,8	0,98834	3,61959	1,01821
64	0,80783	2,27538	1,17318	84,0	0,98907	3,65186	1,01724
65	0,82139	2,30879	1,16383	84,2	0,98979	3,68525	1,01628
66	0,83457	2,34390	1,15455	84,4	0,99048	3,71984	1,01534
67	0,84733	2,38087	1,14535	84,6	0,99114	3,75572	1,01443
68	0,85967	2,41984	1,13624	84,8	0,99178	3,79298	1,01354
69	0,87157	2,46100	1,12725	85,0	0,99240	3,83174	1,01266
70,0	0,88302	2,50455	1,11838	85,2	0,99300	3,87211	1,01181
70,5	0,88857	2,52729	1,11399	85,4	0,99357	3,91423	1,01099
71,0	0,89401	2,55073	1,10964	85,6	0,99411	3,95827	1,01018
71,5	0,89932	2,57490	1,10533	85,8	0,99464	4,00437	1,00940
72,0	0,90451	2,59982	1,10106	86,0	0,99513	4,05276	1,00865
72,5	0,90958	2,62555	1,09683	86,2	0,99561	4,10366	1,00792
73,0	0,91452	2,65214	1,09265	86,4	0,99606	4,15736	1,00721
73,5	0,91934	2,67962	1,08851	86,6	0,99648	4,21416	1,00653
74,0	0,92402	2,70807	1,08443	86,8	0,99688	4,27444	1,00588
74,5	0,92858	2,73752	1,08039	87,0	0,99726	4,33865	1,00526
75,0	0,93301	2,76806	1,07641	87,2	0,99761	4,40733	1,00466
75,5	0,93731	2,79975	1,07248	87,4	0,99794	4,48115	1,00410
76,0	0,94147	2,83267	1,06861	87,6	0,99825	4,56090	1,00356
76,5	0,94550	2,86691	1,06480	87,8	0,99854	4,64765	1,00306
77,0	0,94940	2,90256	1,06106	88,0	0,99878	4,74272	1,00258
77,5	0,95315	2,93974	1,05738	88,2	0,99901	4,84785	1,00215
78,0	0,95677	2,97857	1,05378	88,4	0,99922	4,96542	1,00174
78,5	0,96025	3,01918	1,05024	88,6	0,99940	5,09876	1,00137
79,0	0,96359	3,06173	1,04679	88,8	0,99956	5,25274	1,00104
79,5	0,96679	3,10640	1,04341	89,0	0,99970	5,43491	1,00075
80,0	0,96985	3,15339	1,04011	89,1	0,99975	5,54020	1,00062
80,2	0,97103	3,17288	1,03882	89,2	0,99981	5,65792	1,00050
80,4	0,97219	3,19280	1,03754	89,3	0,99985	5,79140	1,00039
80,6	0,97332	3,21317	1,03628	89,4	0,99989	5,94550	1,00030
80,8	0,97444	3,23400	1,03503	89,5	0,99992	6,12778	1,00021
81,0	0,97553	3,25530	1,03379	89,6	0,99995	6,35088	1,00014
81,2	0,97660	3,27711	1,03257	89,7	0,99997	6,63854	1,00008
81,4	0,97764	3,29945	1,03136	89,8	0,99999	7,04398	1,00004
81,6	0,97866	3,32234	1,03017	89,9	1,00000	7,73711	1,00001
81,8	0,97966	3,34580	1,02900	90,0	1,00000	∞	1,00000
82,0	0,98063	3,36987	1,02784				
82,2	0,98158	3,39457	1,02670				

h^2	K	K'	K'/K	K/K'	$\ln h$	$\ln h'$	k'^2
0,00	1,57080	∞	∞	0,00000	−∞	0,00000	1,00
0,01	1,57475	3,69564	2,34682	0,42611	0,79806-4	0,41863-1	0,99
0,02	1,57874	3,35411	2,12457	0,47068	0,10129-3	0,35781-1	0,98
0,03	1,58278	3,15587	1,99388	0,50153	0,27960-3	0,31572-1	0,97
0,04	1,58687	3,01611	1,90067	0,52613	0,40677-3	0,28216-1	0,96
0,05	1,59100	2,90834	1,82799	0,54705	0,50393-3	0,25362-1	0,95
0,06	1,59519	2,82075	1,76828	0,56552	0,58738-3	0,22842-1	0,94
0,07	1,59942	2,74707	1,71754	0,58223	0,65663-3	0,20562-1	0,93
0,08	1,60371	2,68355	1,67334	0,59761	0,71693-3	0,18464-1	0,92
0,09	1,60805	2,62777	1,63414	0,61194	0,77042-3	0,16508-1	0,91
0,10	1,61244	2,57809	1,59887	0,62544	0,81853-3	0,14666-1	0,90
0,11	1,61689	2,53333	1,56680	0,63825	0,86230-3	0,12919-1	0,89
0,12	1,62139	2,49264	1,53734	0,65047	0,90249-3	0,11251-1	0,88
0,13	1,62595	2,45534	1,51009	0,66221	0,93967-3	0,09649-1	0,87
0,14	1,63058	2,42093	1,48471	0,67353	0,97430-3	0,08105-1	0,86
0,15	1,63526	2,38902	1,46094	0,68449	0,00672-2	0,06610-1	0,85
0,16	1,64000	2,35926	1,43858	0,69513	0,03724-2	0,05158-1	0,84
0,17	1,64481	2,33141	1,41744	0,70550	0,06608-2	0,03743-1	0,83
0,18	1,64968	2,30523	1,39738	0,71562	0,09344-2	0,02362-1	0,82
0,19	1,65462	2,28055	1,37829	0,72553	0,11949-2	0,01010-1	0,81
0,20	1,65962	2,25721	1,36007	0,73526	0,14435-2	0,99683-2	0,80
0,21	1,66470	2,23507	1,34262	0,74481	0,16816-2	0,98380-2	0,79
0,22	1,66985	2,21402	1,32588	0,75422	0,19099-2	0,97097-2	0,78
0,23	1,67507	2,19397	1,30978	0,76349	0,21297-2	0,95831-2	0,77
0,24	1,68037	2,17483	1,29425	0,77265	0,23415-2	0,94582-2	0,76
0,25	1,68575	2,15652	1,27926	0,78171	0,25461-2	0,93347-2	0,75
0,26	1,69121	2,13897	1,26476	0,79066	0,27439-2	0,92124-2	0,74
0,27	1,69675	2,12213	1,25070	0,79955	0,29356-2	0,90911-2	0,73
0,28	1,70237	2,10595	1,23707	0,80836	0,31218-2	0,89709-2	0,72
0,29	1,70809	2,09037	1,22381	0,81712	0,33026-2	0,88514-2	0,71
0,30	1,71389	2,07536	1,21091	0,82583	0,34787-2	0,87326-2	0,70
0,31	1,71978	2,06088	1,19834	0,83449	0,36502-2	0,86144-2	0,69
0,32	1,72577	2,04689	1,18607	0,84312	0,38175-2	0,84967-2	0,68
0,33	1,73186	2,03336	1,17409	0,85172	0,39810-2	0,83793-2	0,67
0,34	1,73805	2,02028	1,16238	0,86030	0,41408-2	0,82622-2	0,66
0,35	1,74435	2,00760	1,15091	0,86887	0,42972-2	0,81453-2	0,65
0,36	1,75075	1,99530	1,13986	0,87744	0,44504-2	0,80284-2	0,64
0,37	1,75727	1,98337	1,12867	0,88600	0,46007-2	0,79116-2	0,63
0,38	1,76390	1,97178	1,11786	0,89457	0,47482-2	0,77947-2	0,62
0,39	1,77065	1,96052	1,10723	0,90315	0,48932-2	0,76776-2	0,61
0,40	1,77752	1,94957	1,09679	0,91175	0,50356-2	0,75603-2	0,60
0,41	1,78452	1,93891	1,08652	0,92037	0,51758-2	0,74426-2	0,59
0,42	1,79165	1,92853	1,07640	0,92903	0,53139-2	0,73246-2	0,58
0,43	1,79892	1,91841	1,06642	0,93771	0,54500-2	0,72061-2	0,57
0,44	1,80633	1,90855	1,05659	0,94644	0,55841-2	0,70870-2	0,56
0,45	1,81388	1,89892	1,04688	0,95522	0,57166-2	0,69673-2	0,55
0,46	1,82159	1,88953	1,03730	0,96404	0,58474-2	0,68468-2	0,54
0,47	1,82946	1,88036	1,02782	0,97293	0,59766-2	0,67256-2	0,53
0,48	1,83749	1,87140	1,01845	0,98188	0,61045-2	0,66035-2	0,52
0,49	1,84569	1,86264	1,00918	0,99090	0,62310-2	0,64804-2	0,51
0,50	1,85407	1,85407	1,00000	1,00000	0,63562-2	0,63562-2	0,50
k'^2	K'	K	K/K'	K'/K	$\ln h'$	$\ln h$	k^2

SMALL VALUES OF THE MODULUS

k^2	K	K'	K'/K	K/K'	h'^2
0,000001	1,57080	8,29405	5,28016	0,18939	0,999999
0,000002	1,57080	7,94748	5,05952	0,19765	0,999998
0,000003	1,57080	7,74475	4,93046	0,20282	0,999997
0,000004	1,57080	7,60091	4,83888	0,20666	0,999996
0,000005	1,57080	7,48934	4,76786	0,20974	0,999995
0,000006	1,57080	7,39818	4,70982	0,21232	0,999994
0,000007	1,57080	7,32111	4,66075	0,21456	0,999993
0,000008	1,57080	7,25434	4,61825	0,21653	0,999992
0,000009	1,57080	7,19545	4,58076	0,21830	0,999991
0,000010	1,57080	7,14277	4,54722	0,21991	0,999990
0,000100	1,57083	5,99159	3,81427	0,26217	0,999900
0,000200	1,57087	5,64512	3,59362	0,27827	0,999800
0,000300	1,57091	5,44249	3,46454	0,28864	0,999700
0,000400	1,57095	5,29875	3,37295	0,29648	0,999600
0,000500	1,57099	5,18727	3,30191	0,30286	0,999500
0,000600	1,57103	5,09620	3,24385	0,30828	0,999400
0,000700	1,57107	5,01921	3,19477	0,31301	0,999300
0,000800	1,57111	4,95253	3,15225	0,31723	0,999200
0,000900	1,57115	4,89373	3,11474	0,32105	0,999100
0,001000	1,57119	4,84113	3,08119	0,32455	0,999000
0,001100	1,57123	4,79356	3,05084	0,32778	0,998900
0,001200	1,57127	4,75014	3,02312	0,33078	0,998800
0,001300	1,57131	4,71020	2,99763	0,33360	0,998700
0,001400	1,57135	4,67322	2,97402	0,33624	0,998600
0,001500	1,57139	4,63880	2,95205	0,33875	0,998500
0,001600	1,57142	4,60661	2,93149	0,34112	0,998400
0,001700	1,57146	4,57638	2,91217	0,34339	0,998300
0,001800	1,57150	4,54788	2,89396	0,34555	0,998200
0,001900	1,57154	4,52092	2,87674	0,34762	0,998100
0,002000	1,57158	4,49535	2,86040	0,34960	0,998000
0,002100	1,57162	4,47103	2,84485	0,35151	0,997900
0,002200	1,57166	4,44784	2,83002	0,35335	0,997800
0,002300	1,57171	4,42569	2,81586	0,35513	0,997700
0,002400	1,57174	4,40448	2,80231	0,35685	0,997600
0,002500	1,57178	4,38414	2,78929	0,35851	0,997500
0,002600	1,57182	4,36461	2,77679	0,36013	0,997400
0,002700	1,57186	4,34581	2,76476	0,36170	0,997300
0,002800	1,57190	4,32769	2,75317	0,36322	0,997200
0,002900	1,57194	4,31022	2,74198	0,36470	0,997100
0,003000	1,57198	4,29334	2,73117	0,36614	0,997000
h'^2	K'	K	K/K'	K'/K	k^2

II. The elliptic integral of the first kind

$$F(\varphi,\ k) = \int_0^\varphi \frac{dt}{\sqrt{1 - k^2 \sin^2 t}}, \quad k = \sin \alpha$$

$\varphi°$	$\alpha = 5°$	$\alpha = 10°$	$\alpha = 15°$	$\alpha = 20°$	$\alpha = 25°$	$\alpha = 30°$
1	0,01745	0,01745	0,01745	0,01745	0,01745	0,01745
2	0,03491	0,03491	0,03491	0,03491	0,03491	0,03491
3	0,05236	0,05236	0,05236	0,05236	0,05236	0,05237
4	0,06981	0,06981	0,06982	0,06982	0,06982	0,06983
5	0,08727	0,08727	0,08728	0,08728	0,08729	0,08729
6	0,1047	0,1047	0,1047	0,1047	0,1048	0,1048
7	0,1222	0,1222	0,1222	0,1222	0,1222	0,1223
8	0,1396	0,1396	0,1397	0,1397	0,1397	0,1397
9	0,1571	0,1571	0,1571	0,1572	0,1572	0,1572
10	0,1745	0,1746	0,1746	0,1746	0,1747	0,1748
11	0,1920	0,1920	0,1921	0,1921	0,1922	0,1923
12	0,2095	0,2095	0,2095	0,2096	0,2097	0,2098
13	0,2269	0,2270	0,2270	0,2271	0,2272	0,2274
14	0,2444	0,2444	0,2445	0,2446	0,2448	0,2450
15	0,2618	0,2619	0,2620	0,2622	0,2623	0,2625
16	0,2793	0,2794	0,2795	0,2797	0,2799	0,2802
17	0,2967	0,2968	0,2970	0,2972	0,2975	0,2978
18	0,3142	0,3143	0,3145	0,3148	0,3151	0,3154
19	0,3317	0,3318	0,3320	0,3323	0,3327	0,3331
20	0,3491	0,3493	0,3495	0,3499	0,3503	0,3508
21	0,3666	0,3668	0,3671	0,3675	0,3680	0,3686
22	0,3840	0,3843	0,3846	0,3851	0,3856	0,3863
23	0,4015	0,4017	0,4021	0,4027	0,4033	0,4041
24	0,4190	0,4192	0,4197	0,4203	0,4210	0,4219
25	0,4364	0,4367	0,4372	0,4379	0,4388	0,4397
26	0,4539	0,4542	0,4548	0,4556	0,4565	0,4576
27	0,4714	0,4717	0,4724	0,4732	0,4743	0,4755
28	0,4888	0,4893	0,4899	0,4909	0,4921	0,4935
29	0,5063	0,5068	0,5075	0,5086	0,5099	0,5114
30	0,5238	0,5243	0,5251	0,5263	0,5277	0,5294
31	0,5412	0,5418	0,5427	0,5440	0,5456	0,5475
32	0,5587	0,5593	0,5604	0,5618	0,5635	0,5656
33	0,5762	0,5769	0,5780	0,5795	0,5814	0,5837
34	0,5937	0,5944	0,5956	0,5973	0,5994	0,6018
35	0,6111	0,6119	0,6133	0,6151	0,6173	0,6200
36	0,6286	0,6295	0,6309	0,6329	0,6355	0,6383
37	0,6461	0,6470	0,6486	0,6507	0,6534	0,6566
38	0,6636	0,6646	0,6662	0,6685	0,6714	0,6749
39	0,6810	0,6821	0,6839	0,6864	0,6895	0,6932
40	0,6985	0,6997	0,7016	0,7043	0,7077	0,7117
41	0,7160	0,7173	0,7193	0,7222	0,7258	0,7301
42	0,7335	0,7348	0,7370	0,7401	0,7440	0,7486
43	0,7510	0,7524	0,7548	0,7581	0,7622	0,7671
44	0,7685	0,7700	0,7725	0,7760	0,7804	0,7857
45	0,7859	0,7876	0,7903	0,7940	0,7987	0,8044

$$F(\varphi, k), \quad k = \sin \alpha$$

$\varphi°$	$\alpha = 5°$	$\alpha = 10°$	$\alpha = 15°$	$\alpha = 20°$	$\alpha = 25°$	$\alpha = 30°$
46	0,8034	0,8052	0,8080	0,8120	0,8170	0,8231
47	0,8209	0,8228	0,8258	0,8300	0,8354	0,8418
48	0,8384	0,8404	0,8436	0,8480	0,8537	0,8606
49	0,8559	0,8580	0,8614	0,8661	0,8721	0,8794
50	0,8734	0,8756	0,8792	0,8842	0,8905	0,8983
51	0,8909	0,8932	0,8970	0,9023	0,9090	0,9172
52	0,9084	0,9108	0,9148	0,9204	0,9275	0,9361
53	0,9259	0,9284	0,9326	0,9385	0,9460	0,9551
54	0,9434	0,9460	0,9505	0,9567	0,9646	0,9742
55	0,9609	0,9637	0,9683	0,9748	0,9832	0,9933
56	0,9784	0,9813	0,9862	0,9930	1,0018	1,0125
57	0,9959	0,9989	1,0041	1,0112	1,0204	1,0317
58	1,0134	1,0166	1,0219	1,0295	1,0391	1,0509
59	1,0309	1,0342	1,0398	1,0477	1,0578	1,0702
60	1,0484	1,0519	1,0577	1,0660	1,0766	1,0896
61	1,0659	1,0695	1,0757	1,0843	1,0953	1,1089
62	1,0834	1,0872	1,0936	1,1026	1,1141	1,1284
63	1,1009	1,1049	1,1115	1,1209	1,1330	1,1478
64	1,1184	1,1225	1,1295	1,1392	1,1518	1,1674
65	1,1359	1,1402	1,1474	1,1576	1,1707	1,1869
66	1,1534	1,1579	1,1654	1,1759	1,1896	1,2065
67	1,1709	1,1756	1,1833	1,1943	1,2085	1,2262
68	1,1884	1,1932	1,2013	1,2127	1,2275	1,2458
69	1,2059	1,2109	1,2193	1,2311	1,2465	1,2656
70	1,2235	1,2286	1,2373	1,2495	1,2655	1,2853
71	1,2410	1,2463	1,2553	1,2680	1,2845	1,3051
72	1,2585	1,2640	1,2733	1,2864	1,3036	1,3249
73	1,2760	1,2817	1,2913	1,3049	1,3226	1,3448
74	1,2935	1,2994	1,3093	1,3234	1,3417	1,3647
75	1,3110	1,3171	1,3273	1,3418	1,3608	1,3846
76	1,3285	1,3348	1,3454	1,3603	1,3800	1,4045
77	1,3461	1,3525	1,3634	1,3788	1,3991	1,4245
78	1,3636	1,3702	1,3814	1,3974	1,4183	1,4445
79	1,3811	1,3879	1,3995	1,4159	1,4374	1,4645
80	1,3986	1,4057	1,4175	1,4344	1,4566	1,4846
81	1,4161	1,4234	1,4356	1,4530	1,4758	1,5046
82	1,4336	1,4411	1,4536	1,4715	1,4950	1,5247
83	1,4512	1,4588	1,4717	1,4901	1,5143	1,5448
84	1,4687	1,4765	1,4897	1,5086	1,5335	1,5649
85	1,4862	1,4942	1,5078	1,5272	1,5527	1,5850
86	1,5037	1,5120	1,5259	1,5457	1,5720	1,6052
87	1,5212	1,5297	1,5439	1,5643	1,5912	1,6253
88	1,5388	1,5474	1,5620	1,5829	1,6105	1,6455
89	1,5563	1,5651	1,5801	1,6015	1,6297	1,6656
90	1,5738	1,5828	1,5981	1,6200	1,6490	1,6858

$$F(\varphi, k), \quad k = \sin \alpha$$

$\varphi°$	$\alpha = 35°$	$\alpha = 40°$	$\alpha = 45°$	$\alpha = 50°$	$\alpha = 55°$	$\alpha = 60°$
1	0,01745	0,01745	0,01745	0,01745	0,01745	0,01745
2	0,03491	0,03491	0,03491	0,03491	0,03491	0,03491
3	0,05237	0,05237	0,05237	0,05237	0,05238	0,05238
4	0,06983	0,06984	0,06984	0,06985	0,06985	0,06986
5	0,08730	0,08731	0,08732	0,08733	0,08734	0,08735
6	0,1048	0,1048	0,1048	0,1048	0,1049	0,1049
7	0,1223	0,1223	0,1223	0,1224	0,1224	0,1224
8	0,1398	0,1398	0,1399	0,1399	0,1399	0,1400
9	0,1573	0,1574	0,1574	0,1575	0,1575	0,1576
10	0,1748	0,1749	0,1750	0,1751	0,1751	0,1752
11	0,1924	0,1925	0,1926	0,1927	0,1928	0,1929
12	0,2099	0,2101	0,2102	0,2103	0,2105	0,2106
13	0,2275	0,2277	0,2279	0,2280	0,2282	0,2284
14	0,2451	0,2454	0,2456	0,2458	0,2460	0,2462
15	0,2628	0,2630	0,2633	0,2636	0,2638	0,2641
16	0,2804	0,2808	0,2811	0,2814	0,2817	0,2820
17	0,2981	0,2985	0,2989	0,2993	0,2997	0,3000
18	0,3159	0,3163	0,3168	0,3172	0,3177	0,3181
19	0,3336	0,3341	0,3347	0,3352	0,3357	0,3362
20	0,3514	0,3520	0,3526	0,3533	0,3539	0,3545
21	0,3692	0,3699	0,3706	0,3714	0,3721	0,3728
22	0,3871	0,3879	0,3887	0,3896	0,3904	0,3912
23	0,4049	0,4059	0,4068	0,4078	0,4088	0,4097
24	0,4229	0,4239	0,4250	0,4261	0,4272	0,4283
25	0,4408	0,4420	0,4433	0,4446	0,4458	0,4470
26	0,4589	0,4602	0,4616	0,4630	0,4645	0,4658
27	0,4769	0,4784	0,4800	0,4816	0,4832	0,4847
28	0,4950	0,4967	0,4985	0,5003	0,5021	0,5038
29	0,5132	0,5150	0,5170	0,5190	0,5210	0,5229
30	0,5313	0,5334	0,5356	0,5379	0,5401	0,5422
31	0,5496	0,5519	0,5543	0,5568	0,5593	0,5617
32	0,5679	0,5704	0,5731	0,5759	0,5786	0,5812
33	0,5862	0,5890	0,5920	0,5950	0,5980	0,6010
34	0,6046	0,6077	0,6109	0,6143	0,6176	0,6208
35	0,6231	0,6264	0,6300	0,6336	0,6373	0,6409
36	0,6416	0,6452	0,6491	0,6531	0,6572	0,6610
37	0,6602	0,6641	0,6684	0,6727	0,6771	0,6814
38	0,6788	0,6831	0,6877	0,6925	0,6973	0,7020
39	0,6975	0,7021	0,7071	0,7123	0,7176	0,7227
40	0,7162	0,7213	0,7267	0,7323	0,7380	0,7436
41	0,7350	0,7405	0,7463	0,7524	0,7586	0,7647
42	0,7539	0,7598	0,7661	0,7727	0,7794	0,7860
43	0,7728	0,7791	0,7859	0,7931	0,8004	0,8075
44	0,7918	0,7986	0,8059	0,8136	0,8215	0,8293
45	0,8109	0,8182	0,8260	0,8343	0,8428	0,8512

$$F(\varphi, k), \quad k = \sin \alpha$$

$\varphi°$	$\alpha = 35°$	$\alpha = 40°$	$\alpha = 45°$	$\alpha = 50°$	$\alpha = 55°$	$\alpha = 60°$
46	0,8300	0,8378	0,8462	0,8552	0,8643	0,8734
47	0,8492	0,8575	0,8666	0,8761	0,8860	0,8959
48	0,8685	0,8773	0,8870	0,8973	0,9079	0,9185
49	0,8878	0,8973	0,9076	0,9186	0,9300	0,9415
50	0,9072	0,9173	0,9283	0,9401	0,9523	0,9647
51	0,9267	0,9374	0,9491	0,9617	0,9748	0,9881
52	0,9462	0,9576	0,9701	0,9835	0,9976	1,0119
53	0,9658	0,9778	0,9912	1,0055	1,0206	1,0359
54	0,9855	0,9982	1,0124	1,0277	1,0437	1,0602
55	1,0052	1,0187	1,0337	1,0500	1,0672	1,0848
56	1,0250	1,0393	1,0552	1,0725	1,0908	1,1097
57	1,0449	1,0600	1,0768	1,0952	1,1147	1,1349
58	1,0648	1,0807	1,0985	1,1180	1,1389	1,1605
59	1,0848	1,1016	1,1204	1,1411	1,1633	1,1864
60	1,1049	1,1226	1,1424	1,1643	1,1879	1,2126
61	1,1250	1,1436	1,1646	1,1877	1,2128	1,2392
62	1,1453	1,1648	1,1869	1,2113	1,2379	1,2661
63	1,1655	1,1860	1,2093	1,2351	1,2633	1,2933
64	1,1859	1,2074	1,2318	1,2591	1,2890	1,3209
65	1,2063	1,2288	1,2545	1,2833	1,3149	1,3489
66	1,2267	1,2503	1,2773	1,3076	1,3411	1,3773
67	1,2472	1,2719	1,3002	1,3321	1,3675	1,4060
68	1,2678	1,2936	1,3233	1,3568	1,3942	1,4351
69	1,2885	1,3154	1,3464	1,3817	1,4212	1,4646
70	1,3092	1,3372	1,3697	1,4068	1,4484	1,4944
71	1,3299	1,3592	1,3931	1,4320	1,4759	1,5246
72	1,3507	1,3812	1,4167	1,4574	1,5036	1,5552
73	1,3716	1,4033	1,4403	1,4830	1,5316	1,5862
74	1,3924	1,4254	1,4640	1,5087	1,5597	1,6175
75	1,4134	1,4477	1,4879	1,5346	1,5882	1,6492
76	1,4344	1,4700	1,5118	1,5608	1,6168	1,6812
77	1,4554	1,4923	1,5359	1,5867	1,6457	1,7136
78	1,4765	1,5147	1,5600	1,6130	1,6748	1,7463
79	1,4976	1,5372	1,5842	1,6394	1,7040	1,7792
80	1,5187	1,5597	1,6085	1,6660	1,7335	1,8125
81	1,5399	1,5823	1,6328	1,6926	1,7631	1,8461
82	1,5611	1,6049	1,6573	1,7194	1,7929	1,8799
83	1,5823	1,6276	1,6817	1,7462	1,8228	1,9140
84	1,6035	1,6502	1,7063	1,7731	1,8528	1,9482
85	1,6248	1,6730	1,7308	1,8001	1,8830	1,9826
86	1,6461	1,6957	1,7554	1,8271	1,9132	2,0172
87	1,6673	1,7184	1,7801	1,8542	1,9435	2,0519
88	1,6886	1,7412	1,8047	1,8813	1,9739	2,0867
89	1,7099	1,7640	1,8294	1,9084	2,0043	2,1216
90	1,7313	1,7868	1,8541	1,9356	2,0347	2,1565

$$F(\varphi, k), \quad k = \sin\alpha$$

$\varphi°$	$\alpha = 65°$	$\alpha = 70°$	$\alpha = 75°$	$\alpha = 80°$	$\alpha = 85°$	$\alpha = 90°$
1	0,01745	0,01745	0,01745	0,01745	0,01745	0,01745
2	0,03491	0,03491	0,03491	0,03491	0,03491	0,03491
3	0,05238	0,05238	0,05238	0,05238	0,05238	0,05238
4	0,06986	0,06986	0,06987	0,06987	0,06987	0,06987
5	0,08736	0,08736	0,08737	0,08737	0,08738	0,08738
6	0,1049	0,1049	0,1049	0,1049	0,1049	0,1049
7	0,1224	0,1224	0,1225	0,1225	0,1225	0,1225
8	0,1400	0,1400	0,1401	0,1401	0,1401	0,1401
9	0,1576	0,1577	0,1577	0,1577	0,1577	0,1577
10	0,1753	0,1753	0,1754	0,1754	0,1754	0,1754
11	0,1930	0,1930	0,1931	0,1931	0,1932	0,1932
12	0,2107	0,2108	0,2109	0,2109	0,2110	0,2110
13	0,2285	0,2286	0,2287	0,2288	0,2289	0,2289
14	0,2464	0,2465	0,2466	0,2467	0,2468	0,2468
15	0,2643	0,2645	0,2646	0,2648	0,2648	0,2648
16	0,2823	0,2825	0,2827	0,2828	0,2829	0,2830
17	0,3003	0,3006	0,3009	0,3010	0,3011	0,3012
18	0,3185	0,3188	0,3191	0,3193	0,3194	0,3195
19	0,3367	0,3371	0,3374	0,3377	0,3378	0,3379
20	0,3550	0,3555	0,3559	0,3562	0,3563	0,3564
21	0,3734	0,3740	0,3744	0,3747	0,3749	0,3750
22	0,3919	0,3926	0,3931	0,3935	0,3937	0,3938
23	0,4105	0,4113	0,4119	0,4123	0,4126	0,4127
24	0,4293	0,4301	0,4308	0,4313	0,4316	0,4317
25	0,4481	0,4490	0,4498	0,4504	0,4508	0,4509
26	0,4670	0,4681	0,4690	0,4697	0,4701	0,4702
27	0,4861	0,4874	0,4884	0,4891	0,4896	0,4897
28	0,5053	0,5067	0,5079	0,5087	0,5092	0,5094
29	0,5247	0,5262	0,5275	0,5285	0,5291	0,5293
30	0,5442	0,5459	0,5474	0,5484	0,5491	0,5493
31	0,5639	0,5658	0,5674	0,5686	0,5693	0,5696
32	0,5837	0,5858	0,5876	0,5889	0,5898	0,5900
33	0,6037	0,6060	0,6080	0,6095	0,6104	0,6107
34	0,6238	0,6265	0,6287	0,6303	0,6313	0,6317
35	0,6442	0,6471	0,6495	0,6513	0,6525	0,6528
36	0,6647	0,6679	0,6706	0,6726	0,6739	0,6743
37	0,6854	0,6890	0,6919	0,6941	0,6955	0,6960
38	0,7063	0,7102	0,7135	0,7159	0,7175	0,7180
39	0,7275	0,7318	0,7353	0,7380	0,7397	0,7403
40	0,7488	0,7535	0,7575	0,7604	0,7623	0,7629
41	0,7704	0,7756	0,7799	0,7831	0,7852	0,7859
42	0,7922	0,7979	0,8026	0,8062	0,8084	0,8092
43	0,8143	0,8205	0,8256	0,8295	0,8320	0,8328
44	0,8367	0,8433	0,8490	0,8533	0,8560	0,8569
45	0,8593	0,8665	0,8727	0,8774	0,8804	0,8814

$$F(\varphi,\ k),\quad k=\sin\alpha$$

$\varphi°$	$\alpha = 65°$	$\alpha = 70°$	$\alpha = 75°$	$\alpha = 80°$	$\alpha = 85°$	$\alpha = 90°$
46	0,8821	0,8901	0,8968	0,9019	0,9052	0,9063
47	0,9053	0,9139	0,9212	0,9269	0,9304	0,9316
48	0,9288	0,9381	0,9461	0,9523	0,9561	0,9575
49	0,9525	0,9627	0,9714	0,9781	0,9824	0,9838
50	0,9766	0,9876	0,9071	1,0044	1,0091	1,0107
51	1,0010	1,0130	1,0233	1,0313	1,0364	1,0381
52	1,0258	1,0387	1,0500	1,0587	1,0643	1,0662
53	1,0509	1,0649	1,0771	1,0867	1,0927	1,0948
54	1,0764	1,0916	1,1048	1,1152	1,1219	1,1242
55	1,1022	1,1187	1,1331	1,1444	1,1517	1,1542
56	1,1285	1,1462	1,1619	1,1743	1,1823	1,1851
57	1,1551	1,1743	1,1914	1,2049	1,2136	1,2167
58	1,1822	1,2030	1,2215	1,2362	1,2458	1,2492
59	1,2097	1,2321	1,2522	1,2684	1,2789	1,2826
60	1,2376	1,2619	1,2837	1,3014	1,3129	1,3170
61	1,2660	1,2922	1,3159	1,3352	1,3480	1,3524
62	1,2949	1,3231	1,3490	1,3701	1,3841	1,3890
63	1,3243	1,3547	1,3828	1,4059	1,4214	1,4268
64	1,3541	1,3870	1,4175	1,4429	1,4599	1,4659
65	1,3844	1,4199	1,4532	1,4810	1,4998	1,5065
66	1,4153	1,4536	1,4898	1,5203	1,5411	1,5485
67	1,4467	1,4880	1,5274	1,5610	1,5840	1,5923
68	1,4786	1,5232	1,5661	1,6030	1,6287	1,6379
69	1,5111	1,5591	1,6059	1,6466	1,6752	1,6856
70	1,5441	1,5959	1,6468	1,6918	1,7237	1,7354
71	1,5777	1,6335	1,6891	1,7388	1,7745	1,7877
72	1,6118	1,6720	1,7326	1,7876	1,8277	1,8427
73	1,6465	1,7113	1,7774	1,8384	1,8837	1,9008
74	1,6818	1,7516	1,8237	1,8915	1,9427	1,9623
75	1,7176	1,7927	1,8715	1,9468	2,0050	2,0276
76	1,7540	1,8347	1,9207	2,0047	2,0711	2,0973
77	1,7909	1,8777	1,9716	2,0653	2,1414	2,1721
78	1,8284	1,9215	2,0240	2,1288	2,2164	2,2528
79	1,8664	1,9663	2,0781	2,1954	2,2969	2,3404
80	1,9048	2,0119	2,1339	2,2653	2,3837	2,4362
81	1,9438	2,0584	2,1913	2,3387	2,4775	2,5421
82	1,9831	2,1057	2,2504	2,4157	2,5795	2,6603
83	2,0229	2,1537	2,3110	2,4965	2,6911	2,7942
84	2,0630	2,2024	2,3731	2,5811	2,8136	2,9487
85	2,1035	2,2518	2,4366	2,6694	2,9487	3,1313
86	2,1442	2,3017	2,5013	2,7612	3,0978	3,3547
87	2,1852	2,3520	2,5670	2,8561	3,2620	3,6425
88	2,2263	2,4027	2,6336	2,9537	3,4412	4,0481
89	2,2675	2,4535	2,7007	3,0530	3,6328	4,7413
90	2,3088	2,5046	2,7681	3,1534	3,8317	∞

III. The elliptic integral of the second kind

$$E(\varphi,\ k)=\int_0^{\varphi}\sqrt{1-k^2\sin^2 t}\,dt,\ k=\sin\alpha$$

$\varphi°$	$\alpha = 5°$	$\alpha = 10°$	$\alpha = 15°$	$\alpha = 20°$	$\alpha = 25°$	$\alpha = 30°$
1	0,0175	0,0175	0,0175	0,0175	0,0175	0,0175
2	0,0349	0,0349	0,0349	0,0349	0,0349	0,0349
3	0,0524	0,0524	0,0524	0,0524	0,0524	0,0524
4	0,0698	0,0698	0,0698	0,0698	0,0698	0,0698
5	0,0873	0,0873	0,0873	0,0873	0,0873	0,0872
6	0,1047	0,1047	0,1047	0,1047	0,1047	0,1047
7	0,1222	0,1222	0,1222	0,1221	0,1221	0,1221
8	0,1396	0,1396	0,1396	0,1396	0,1396	0,1395
9	0,1571	0,1571	0,1570	0,1570	0,1570	0,1569
10	0,1745	0,1745	0,1745	0,1744	0,1744	0,1743
11	0,1920	0,1920	0,1919	0,1919	0,1918	0,1917
12	0,2094	0,2094	0,2093	0,2093	0,2092	0,2091
13	0,2269	0,2268	0,2268	0,2267	0,2266	0,2264
14	0,2443	0,2443	0,2442	0,2441	0,2439	0,2437
15	0,2618	0,2617	0,2616	0,2615	0,2613	0,2611
16	0,2792	0,2791	0,2790	0,2788	0,2786	0,2784
17	0,2967	0,2966	0,2964	0,2962	0,2959	0,2956
18	0,3141	0,3140	0,3138	0,3136	0,3133	0,3129
19	0,3316	0,3314	0,3312	0,3309	0,3305	0,3301
20	0,3490	0,3489	0,3486	0,3483	0,3478	0,3473
21	0,3665	0,3663	0,3660	0,3656	0,3651	0,3645
22	0,3839	0,3837	0,3834	0,3829	0,3823	0,3817
23	0,4014	0,4011	0,4007	0,4002	0,3996	0,3988
24	0,4188	0,4185	0,4181	0,4175	0,4168	0,4159
25	0,4362	0,4359	0,4354	0,4348	0,4339	0,4330
26	0,4537	0,4533	0,4528	0,4520	0,4511	0,4500
27	0,4711	0,4707	0,4701	0,4693	0,4682	0,4670
28	0,4886	0,4881	0,4875	0,4865	0,4854	0,4840
29	0,5060	0,5055	0,5048	0,5037	0,5025	0,5010
30	0,5234	0,5229	0,5221	0,5209	0,5195	0,5179
31	0,5409	0,5403	0,5394	0,5381	0,5366	0,5348
32	0,5583	0,5577	0,5567	0,5553	0,5536	0,5516
33	0,5757	0,5751	0,5740	0,5725	0,5706	0,5684
34	0,5932	0,5924	0,5912	0,5896	0,5876	0,5852
35	0,6106	0,6098	0,6085	0,6067	0,6045	0,6019
36	0,6280	0,6272	0,6258	0,6238	0,6214	0,6186
37	0,6455	0,6445	0,6430	0,6409	0,6383	0,6353
38	0,6629	0,6619	0,6602	0,6580	0,6552	0,6519
39	0,6803	0,6792	0,6775	0,6750	0,6720	0,6685
40	0,6977	0,6966	0,6947	0,6921	0,6888	0,6851
41	0,7152	0,7139	0,7119	0,7091	0,7056	0,7016
42	0,7326	0,7313	0,7291	0,7261	0,7224	0,7180
43	0,7500	0,7486	0,7463	0,7431	0,7391	0,7345
44	0,7674	0,7659	0,7634	0,7600	0,7558	0,7509
45	0,7849	0,7832	0,7806	0,7770	0,7725	0,7672

$$E\,(\varphi,\,k),\ \ k = \sin\alpha$$

$\varphi°$	$\alpha = 5°$	$\alpha = 10°$	$\alpha = 15°$	$\alpha = 20°$	$\alpha = 25°$	$\alpha = 30°$
46	0,8023	0,8006	0,7978	0,7939	0,7891	0,7835
47	0,8197	0,8179	0,8149	0,8108	0,8057	0,7998
48	0,8371	0,8352	0,8320	0,8277	0,8223	0,8160
49	0,8545	0,8525	0,8491	0,8446	0,8389	0,8322
50	0,8719	0,8698	0,8663	0,8614	0,8554	0,8483
51	0,8894	0,8871	0,8834	0,8783	0,8719	0,8644
52	0,9068	0,9044	0,9005	0,8951	0,8884	0,8805
53	0,9242	0,9217	0,9175	0,9119	0,9048	0,8965
54	0,9416	0,9390	0,9346	0,9287	0,9212	0,9125
55	0,9590	0,9562	0,9517	0,9454	0,9376	0,9284
56	0,9764	0,9735	0,9687	0,9622	0,9540	0,9443
57	0,9938	0,9908	0,9858	0,9789	0,9703	0,9602
58	1,0112	1,0080	1,0028	0,9956	0,9866	0,9760
59	1,0286	1,0253	1,0198	1,0123	1,0029	0,9918
60	1,0460	1,0426	1,0368	1,0290	1,0192	1,0076
61	1,0634	1,0598	1,0538	1,0456	1,0354	1,0233
62	1,0808	1,0771	1,0708	1,0623	1,0516	1,0390
63	1,0982	1,0943	1,0878	1,0789	1,0678	1,0546
64	1,1156	1,1115	1,1048	1,0955	1,0839	1,0702
65	1,1330	1,1288	1,1218	1,1121	1,1001	1,0858
66	1,1504	1,1460	1,1387	1,1287	1,1162	1,1013
67	1,1678	1,1632	1,1557	1,1453	1,1323	1,1168
68	1,1852	1,1805	1,1726	1,1619	1,1483	1,1323
69	1,2026	1,1977	1,1896	1,1784	1,1644	1,1478
70	1,2200	1,2149	1,2065	1,1949	1,1804	1,1632
71	1,2374	1,2321	1,2234	1,2115	1,1964	1,1786
72	1,2548	1,2494	1,2403	1,2280	1,2124	1,1939
73	1,2722	1,2666	1,2573	1,2445	1,2284	1,2093
74	1,2896	1,2838	1,2742	1,2609	1,2443	1,2246
75	1,3070	1,3010	1,2911	1,2774	1,2603	1,2399
76	1,3244	1,3182	1,3080	1,2939	1,2762	1,2552
77	1,3418	1,3354	1,3249	1,3104	1,2921	1,2704
78	1,3592	1,3526	1,3417	1,3268	1,3080	1,2857
79	1,3765	1,3698	1,3586	1,3433	1,3239	1,3009
80	1,3939	1,3870	1,3755	1,3597	1,3398	1,3161
81	1,4113	1,4042	1,3924	1,3761	1,3556	1,3312
82	1,4287	1,4214	1,4093	1,3925	1,3715	1,3464
83	1,4461	1,4386	1,4261	1,4090	1,3873	1,3616
84	1,4635	1,4558	1,4430	1,4254	1,4032	1,3767
85	1,4809	1,4729	1,4599	1,4418	1,4190	1,3919
86	1,4983	1,4901	1,4767	1,4582	1,4348	1,4070
87	1,5157	1,5073	1,4936	1,4746	1,4507	1,4221
88	1,5330	1,5245	1,5104	1,4910	1,4665	1,4372
89	1,5504	1,5417	1,5273	1,5074	1,4823	1,4524
90	1,5678	1,5589	1,5442	1,5238	1,4981	1,4675

$$E(\varphi, k), \quad k = \sin \alpha$$

$\varphi°$	$\alpha = 35°$	$\alpha = 40°$	$\alpha = 45°$	$\alpha = 50°$	$\alpha = 55°$	$\alpha = 60°$
1	0,0175	0,0175	0,0175	0,0175	0,0175	0,0175
2	0,0349	0,0349	0,0349	0,0349	0,0349	0,0349
3	0,0524	0,0524	0,0524	0,0524	0,0523	0,0523
4	0,0698	0,0698	0,0698	0,0698	0,0698	0,0698
5	0,0872	0,0872	0,0872	0,0872	0,0872	0,0872
6	0,1047	0,1046	0,1046	0,1046	0,1046	0,1046
7	0,1221	0,1221	0,1220	0,1220	0,1220	0,1220
8	0,1395	0,1394	0,1394	0,1394	0,1393	0,1393
9	0,1569	0,1568	0,1568	0,1567	0,1567	0,1566
10	0,1743	0,1742	0,1741	0,1740	0,1739	0,1739
11	0,1916	0,1915	0,1914	0,1913	0,1912	0,1911
12	0,2089	0,2088	0,2087	0,2086	0,2084	0,2083
13	0,2263	0,2261	0,2259	0,2258	0,2256	0,2254
14	0,2436	0,2434	0,2431	0,2429	0,2427	0,2425
15	0,2608	0,2606	0,2603	0,2601	0,2598	0,2596
16	0,2781	0,2778	0,2775	0,2771	0,2768	0,2766
17	0,2953	0,2949	0,2946	0,2942	0,2938	0,2935
18	0,3125	0,3121	0,3116	0,3112	0,3107	0,3103
19	0,3297	0,3291	0,3286	0,3281	0,3276	0,3271
20	0,3468	0,3462	0,3456	0,3450	0,3444	0,3438
21	0,3639	0,3632	0,3625	0,3618	0,3611	0,3604
22	0,3809	0,3802	0,3793	0,3785	0,3777	0,3770
23	0,3980	0,3971	0,3961	0,3952	0,3943	0,3935
24	0,4150	0,4139	0,4129	0,4118	0,4108	0,4098
25	0,4319	0,4308	0,4296	0,4284	0,4272	0,4261
26	0,4488	0,4475	0,4462	0,4449	0,4436	0,4423
27	0,4657	0,4643	0,4628	0,4613	0,4598	0,4584
28	0,4825	0,4809	0,4793	0,4776	0,4760	0,4744
29	0,4993	0,4975	0,4957	0,4938	0,4920	0,4903
30	0,5161	0,5141	0,5121	0,5100	0,5080	0,5061
31	0,5328	0,5306	0,5283	0,5261	0,5239	0,5218
32	0,5494	0,5470	0,5446	0,5421	0,5396	0,5373
33	0,5660	0,5634	0,5607	0,5580	0,5553	0,5528
34	0,5826	0,5797	0,5768	0,5738	0,5709	0,5681
35	0,5991	0,5960	0,5928	0,5895	0,5863	0,5833
36	0,6155	0,6122	0,6087	0,6052	0,6017	0,5984
37	0,6319	0,6283	0,6245	0,6207	0,6169	0,6134
38	0,6483	0,6444	0,6403	0,6361	0,6321	0,6282
39	0,6646	0,6604	0,6559	0,6515	0,6471	0,6429
40	0,6808	0,6763	0,6715	0,6667	0,6620	0,6575
41	0,6970	0,6921	0,6870	0,6819	0,6768	0,6719
42	0,7132	0,7079	0,7025	0,6969	0,6914	0,6862
43	0,7293	0,7237	0,7178	0,7118	0,7059	0,7003
44	0,7453	0,7393	0,7330	0,7267	0,7204	0,7144
45	0,7613	0,7549	0,7482	0,7414	0,7347	0,7282

$$E\,(\varphi,\ k),\ \ k=\sin\alpha$$

$\varphi°$	$\alpha = 35°$	$\alpha = 40°$	$\alpha = 45°$	$\alpha = 50°$	$\alpha = 55°$	$\alpha = 60°$
46	0,7772	0,7704	0,7633	0,7560	0,7488	0,7420
47	0,7931	0,7858	0,7782	0,7705	0,7629	0,7555
48	0,8089	0,8012	0,7931	0,7849	0,7768	0,7690
49	0,8247	0,8165	0,8079	0,7992	0,7905	0,7823
50	0,8404	0,8317	0,8227	0,8134	0,8042	0,7954
51	0,8560	0,8469	0,8373	0,8275	0,8177	0,8064
52	0,8716	0,8620	0,8518	0,8414	0,8311	0,8212
53	0,8872	0,8770	0,8663	0,8553	0,8444	0,8339
54	0,9026	0,8919	0,8806	0,8690	0,8575	0,8464
55	0,9181	0,9068	0,8949	0,8827	0,8705	0,8588
56	0,9335	0,9216	0,9091	0,8962	0,8834	0,8710
57	0,9488	0,9363	0,9232	0,9097	0,8961	0,8831
58	0,9641	0,9510	0,9372	0,9230	0,9088	0,8950
59	0,9793	0,9656	0,9511	0,9362	0,9213	0,9068
60	0,9945	0,9801	0,9650	0,9403	0,0330	0,0184
61	1,0096	0,9946	0,9787	0,9623	0,9459	0,9299
62	1,0247	1,0090	0,9924	0,9752	0,9580	0,9412
63	1,0397	1,0233	1,0060	0,9880	0,9700	0,9524
64	1,0547	1,0376	1,0195	1,0007	0,9818	0,9634
65	1,0696	1,0518	1,0329	1,0133	0,9936	0,9743
66	1,0845	1,0660	1,0463	1,0259	1,0052	0,9850
67	1,0993	1,0801	1,0596	1,0383	1,0167	0,9958
68	1,1141	1,0941	1,0728	1,0506	1,0282	1,0061
69	1,1289	1,1081	1,0859	1,0628	1,0395	1,0164
70	1,1436	1,1221	1,0990	1,0750	1,0506	1,0266
71	1,1583	1,1359	1,1120	1,0871	1,0617	1,0367
72	1,1729	1,1498	1,1250	1,0991	1,0727	1,0467
73	1,1875	1,1636	1,1379	1,1110	1,0836	1,0565
74	1,2021	1,1773	1,1507	1,1228	1,0944	1,0662
75	1,2167	1,1910	1,1635	1,1346	1,1051	1,0759
76	1,2312	1,2047	1,1762	1,1463	1,1158	1,0854
77	1,2457	1,2183	1,1889	1,1580	1,1263	1,0948
78	1,2601	1,2319	1,2015	1,1695	1,1368	1,1041
79	1,2746	1,2454	1,2141	1,1811	1,1472	1,1133
80	1,2890	1,2590	1,2266	1,1926	1,1576	1,1225
81	1,3034	1,2725	1,2391	1,2040	1,1678	1,1316
82	1,3177	1,2859	1,2516	1,2154	1,1781	1,1406
83	1,3321	1,2994	1,2640	1,2267	1,1883	1,1495
84	1,3464	1,3128	1,2765	1,2381	1,1984	1,1584
85	1,3608	1,3262	1,2889	1,2493	1,2085	1,1673
86	1,3751	1,3396	1,3012	1,2606	1,2186	1,1761
67	1,3894	1,3530	1,3136	1,2719	1,2286	1,1848
88	1,4037	1,3664	1,3260	1,2831	1,2387	1,1936
89	1,4180	1,3798	1,3383	1,2943	1,2487	1,2023
90	1,4323	1,3931	1,3506	1,3055	1,2587	1,2111

$$E(\varphi, k), \quad k = \sin \alpha$$

$\varphi°$	$\alpha = 65°$	$\alpha = 70°$	$\alpha = 75°$	$\alpha = 80°$	$\alpha = 85°$	$\alpha = 90°$
1	0,0175	0,0175	0,0175	0,0175	0,0175	0,0175
2	0,0349	0,0349	0,0349	0,0349	0,0349	0,0349
3	0,0523	0,0523	0,0523	0,0523	0,0523	0,0523
4	0,0698	0,0698	0,0698	0,0698	0,0698	0,0698
5	0,0872	0,0872	0,0872	0,0872	0,0872	0,0872
6	0,1046	0,1046	0,1045	0,1045	0,1045	0,1045
7	0,1219	0,1219	0,1219	0,1219	0,1219	0,1219
8	0,1393	0,1392	0,1392	0,1392	0,1392	0,1392
9	0,1566	0,1565	0,1565	0,1565	0,1564	0,1564
10	0,1738	0,1738	0,1737	0,1737	0,1737	0,1737
11	0,1910	0,1910	0,1909	0,1908	0,1908	0,1908
12	0,2082	0,2081	0,2080	0,2080	0,2079	0,2079
13	0,2253	0,2252	0,2251	0,2250	0,2250	0,2250
14	0,2424	0,2422	0,2421	0,2420	0,2419	0,2419
15	0,2594	0,2592	0,2590	0,2589	0,2588	0,2588
16	0,2763	0,2761	0,2759	0,2758	0,2757	0,2756
17	0,2932	0,2929	0,2927	0,2925	0,2924	0,2924
18	0,3100	0,3096	0,3094	0,3092	0,3091	0,3090
19	0,3267	0,3263	0,3260	0,3258	0,3256	0,3256
20	0,3433	0,3429	0,3425	0,3422	0,3421	0,3420
21	0,3599	0,3593	0,3589	0,3586	0,3584	0,3584
22	0,3763	0,3757	0,3753	0,3749	0,3747	0,3746
23	0,3927	0,3920	0,3915	0,3911	0,3908	0,3907
24	0,4090	0,4082	0,4076	0,4071	0,4068	0,4067
25	0,4251	0,4243	0,4236	0,4230	0,4227	0,4226
26	0,4412	0,4402	0,4394	0,4389	0,4385	0,4384
27	0,4572	0,4561	0,4552	0,4545	0,4541	0,4540
28	0,4730	0,4718	0,4708	0,4701	0,4696	0,4695
29	0,4888	0,4874	0,4863	0,4855	0,4850	0,4848
30	0,5044	0,5029	0,5017	0,5007	0,5002	0,5000
31	0,5199	0,5182	0,5169	0,5159	0,5153	0,5150
32	0,5352	0,5334	0,5319	0,5308	0,5302	0,5299
33	0,5505	0,5485	0,5468	0,5456	0,5449	0,5446
34	0,5656	0,5634	0,5616	0,5603	0,5595	0,5592
35	0,5806	0,5782	0,5762	0,5748	0,5739	0,5736
36	0,5954	0,5928	0,5907	0,5891	0,5881	0,5878
37	0,6101	0,6073	0,6050	0,6032	0,6022	0,6018
38	0,6247	0,6216	0,6191	0,6172	0,6161	0,6157
39	0,6391	0,6357	0,6330	0,6310	0,6297	0,6293
40	0,6533	0,6497	0,6468	0,6446	0,6432	0,6428
41	0,6675	0,6636	0,6604	0,6580	0,6566	0,6561
42	0,6814	0,6772	0,6738	0,6712	0,6697	0,6691
43	0,6952	0,6907	0,6870	0,6843	0,6826	0,6820
44	0,7088	0,7040	0,7001	0,6971	0,6953	0,6947
45	0,7223	0,7172	0,7129	0,7097	0,7078	0,7071

$$E(\varphi, k), \quad k = \sin \alpha$$

$\varphi°$	$\alpha = 65°$	$\alpha = 70°$	$\alpha = 75°$	$\alpha = 80°$	$\alpha = 85°$	$\alpha = 90°$
46	0,7356	0,7301	0,7255	0,7222	0,7201	0,7193
47	0,7488	0,7429	0,7380	0,7344	0,7321	0,7314
48	0,7618	0,7555	0,7503	0,7464	0,7440	0,7431
49	0,7746	0,7679	0,7623	0,7582	0,7556	0,7547
50	0,7872	0,7801	0,7741	0,7697	0,7670	0,7660
51	0,7997	0,7921	0,7858	0,7811	0,7781	0,7772
52	0,8120	0,8039	0,7972	0,7922	0,7891	0,7880
53	0,8242	0,8155	0,8084	0,8031	0,7998	0,7986
54	0,8361	0,8270	0,8194	0,8137	0,8102	0,8090
55	0,8479	0,8382	0,8302	0,8242	0,8204	0,8192
56	0,8595	0,8493	0,8408	0,8344	0,8304	0,8290
57	0,8709	0,8601	0,8511	0,8443	0,8401	0,8387
58	0,8822	0,8707	0,8612	0,8540	0,8496	0,8481
59	0,8933	0,8812	0,8711	0,8635	0,8588	0,8572
60	0,9042	0,8914	0,8808	0,8728	0,8677	0,8660
61	0,9149	0,9015	0,8903	0,8818	0,8764	0,8746
62	0,9254	0,9113	0,8995	0,8905	0,8849	0,8830
63	0,9358	0,9210	0,9085	0,8990	0,8930	0,8910
64	0,9460	0,9304	0,9173	0,9072	0,9009	0,8988
65	0,9561	0,9397	0,9258	0,9152	0,9086	0,9063
66	0,9659	0,9487	0,9341	0,9230	0,9160	0,9136
67	0,9756	0,9576	0,9422	0,9305	0,9231	0,9205
68	0,9852	0,9662	0,9501	0,9377	0,9299	0,9272
69	0,9946	0,9747	0,9578	0,9447	0,9364	0,9336
70	1,0038	0,9830	0,9652	0,9514	0,9427	0,9397
71	1,0129	0,9911	0,9724	0,9579	0,9487	0,9455
72	1,0218	0,9990	0,9794	0,9642	0,9544	0,9511
73	1,0306	1,0067	0,9862	0,9702	0,9599	0,9563
74	1,0392	1,0143	0,9928	0,9759	0,9650	0,9613
75	1,0477	1,0217	0,9992	0,9814	0,9699	0,9659
76	1,0561	1,0290	1,0053	0,9867	0,9745	0,9703
77	1,0643	1,0361	1,0113	0,9917	0,9789	0,9744
78	1,0725	1,0430	1,0171	0,9965	0,9829	0,9782
79	1,0805	1,0498	1,0228	1,0011	0,9867	0,9816
80	1,0884	1,0565	1,0282	1,0054	0,9902	0,9848
81	1,0962	1,0630	1,0335	1,0096	0,9935	0,9877
82	1,1040	1,0695	1,0387	1,0135	0,9965	0,9903
83	1,1116	1,0758	1,0437	1,0173	0,9992	0,9926
84	1,1192	1,0821	1,0486	1,0209	1,0017	0,9945
85	1,1267	1,0883	1,0534	1,0244	1,0039	0,9962
86	1,1342	1,0944	1,0581	1,0277	1,0060	0,9976
87	1,1417	1,1004	1,0628	1,0309	1,0078	0,9986
88	1,1491	1,1064	1,0674	1,0340	1,0095	0,9994
89	1,1565	1,1124	1,0719	1,0371	1,0111	0,9999
90	1,1638	1,1184	1,0764	1,0401	1,0127	1,0000

Bibliography

1. Adolf Hurwitz and R. Courant *Vorlesungen über allgemeinen Funktionentheorie und elliptische Funktionen*, 3rd ed., Springer-Verlag, 1929; reprint, Interscience, 1944; 4th ed., 1964.

2. M. A. Lavrent'ev and B. V. Shabat *Methods of the theory of functions of a complex variable*, 3rd ed.,"Nauka", Moscow, 1965; German transl., VEB Deutscher Verlag Wiss., Berlin, 1967.

3. S. Stoïlow *Theory of functions of a complex variable*. Vols. 1, 2, Ed. Acad. Repub. Pop. Romîne, Bucharest, 1954, 1958. (Romainian)

4. L. I. Volkovyskiĭ, G. L. Lunts, and I. G. Aramanovich *Problem book in the theory of functions of a complex variable*, Fizmatgiz,Moscow, 1961; English transl., Pergamon Press, Oxford, and Addison-Wesley, Reading, Mass., 1965.

5. P. Appell and E. Lacour *Principes de la théorie des fonctions elliptiques et applications*, 2nd ed., Gauthier-Villars, Paris, 1922.

6. Jules Tannery and Jules Monk *Éléments de la théorie des fonctions elliptiques*. Vols. I–IV, Gauthier-Villars, Paris, 1893–1902, reprint, Chelsea, New York, 1972.

7. Werner von Koppenfels and Friedemann Stallmann *Praxis der konformen Abbildungen*, Springer-Verlag, 1959.

8. Eugene Jahnke, Fritz Emde, and Friedrich Lösch *Tables of higher functions*, 6th rev. ed., McGraw-Hill, New York, 1960; Teubner, Stuttgart, 1966.

9. A. M. Zhuravskiĭ *Handbook of elliptic functions*, Izdat. Akad. Nauk SSSR, 1941. (Russian)

10. L. M. Milne-Thomson *Die elliptischen Funktionen von Jacobi*, Springer-Verlag, 1931.

11. B. I. Segal and K. A. Semendyaev *Five-place mathematical tables*, 3rd ed., Fizmatgiz, Moscow, 1962. (Russian)

12. V. M. Belyakov, R. I. Kravtsova, and M. G. Rappoport *Tables of elliptic integrals*. Vols. 1, 2, Izdat. Akad. Nauk SSSR, 1962, 1963; English transl. of Vol. 1, Macmillan, 1965.

13. Ts. D. Lomkatsi *Tables of the Weierstrass elliptic functions*, Vychisl. Tsentr Akad. Nauk SSSR, Moscow, 1967. (Russian)

14. Fritz Oberhettinger and Wilhelm Magnus *Anwendung der elliptischen Funktionen in Physik und Technik*, Springer-Verlag, 1949.

15. L. I. Sedov *Two-dimensional problems in hydrodynamics and aerodynamics*, 2nd ed., "Nauka", Moscow, 1966; English transl. of 1st ed., Interscience, 1965.

16. *Tables of Chebyshev polynomials $S_n(x)$ and $C_n(x)$*, Nat. Bureau of Standards Appl. Math. Ser., vol. 9, U.S. Govt. Printing Office, Washington, DC, 1952.

17. Wilhelm Cauer *Theorie der linearen Wechselstromschaltungen*, 2nd ed., Akademie-Verlag, Berlin, 1954; English transl., *Synthesis of linear communication networks*, McGraw-Hill, 1958.

18. V. A. Taft *Fundamentals of the methods of computing linear electric circuits from their given frequency characteristics*, Izdat. Akad. Nauk SSSR, Moscow, 1954. (Russian)

19. N. I. Akhiezer *Lectures on the theory of approximation*, 2nd ed., "Nauka", Moscow, 1965; German transl., Akademie-Verlag, Berlin, 1967 ; English transl. of 1st ed., Ungar, New York, 1956.

20. E. I. Zolotarev *Application of elliptic functions to questions of functions deviating least and most from zero*, Zap. Imp. Akad. Nauk St. Petersburg, **30** (1877), no. 5; reprinted in his *Collected works* Vol. 2, Izdat. Akad. Nauk SSSR, Moscow, 1932, pp. 1–59. (Russian) Ibuch Fortschritte Math. **9**, 343.

21. N. I. Akhiezer *Aerodynamical investigations*, Ukraïn. Akad. Nauk Trudi Fiz.-Mat. Vīddīlu **7** (1927/28), no. 2. (Ukrainian)

22. ____, *On a problem of E. I. Zolotarev*, Izv. Akad. Nauk SSSR Otdel. Fiz.-Mat. Nauk **1929**, 919–931. (Russian)

23. ____, *Über eine extremale Eigenschaft reationaler Funktionen*, Soobshch. Khar′kov. Mat. Obshch. i Ukrain. Nauchno-Issled. Inst. Mat. Mekh. (4) **6** (1933), 39–45.

24. ____, *Bemerkungen über extremale Eigenschaften einiger mit der Transformation der elliptischen Funktionen zusammenhängender Brüche*, Soobshch. Khar′kov. Mat. Obshch. i Nauchno-Issled. Inst. Math. Mekh. Khar′kov. Gos. Univ. (4) **11** (1935), 27–34.

25. ____, *Über eine Eigenschaft der "elliptischen" Polynome*, Soobshch. Khar′kov. Mat. Obshch. i Nauchno-Issled. Inst. Math. Mekh. Khar′kov. Gos. Univ. (4) **9** (1934), 3–8.

26. ____, *Verallgemeinerung einer Korkine-Zolotareffschen Minimum-Aufgabe*, Soobshch. Nauchno-Issled. Inst. Mat. Mekh. Khar′kov. Gos. Univ. i Khar′kov. Mat. Obshch. (4) **13** (1936), 3–14.

27. ____, *Orthogonal polynomials on several intervals*, Dokl. Akad. Nauk SSSR **134** (1960), 9–12; English transl. in Soviet Math. Dokl. **1** (1960).

28. N. I. Akhiezer and Yu. Ya. Tomchuk *On the theory of orthogonal polynomials over several intervals*, Dokl. Akad. Nauk SSSR **138** (1961), 743–746; English transl. in Soviet Math. Dokl. **2** (1961).

29. Yu. Ya. Tomchuk *Orthogonal polynomials over a system of intervals on the number line*, Khar′kov. Gos. Univ. Uchen. Zap. 135 = Zap. Fiz.-Mat. Fak. i Khar′kov. Mat. Obshch. (4) **29** (1964), 93–128. (Russian)

Subject Index